MW01674509

# Discrete Mathematics

# Discrete Mathematics

Carlos Davenport

NY Research Press
New York

Published by NY Research Press
118-35 Queens Blvd., Suite 400,
Forest Hills, NY 11375, USA
www.nyresearchpress.com

Discrete Mathematics
Carlos Davenport

© 2021 NY Research Press

International Standard Book Number: 978-1-63238-878-0 (Hardback)

This book contains information obtained from authentic and highly regarded sources. All chapters are published with permission under the Creative Commons Attribution Share Alike License or equivalent. A wide variety of references are listed. Permissions and sources are indicated; for detailed attributions, please refer to the permissions page. Reasonable efforts have been made to publish reliable data and information, but the authors, editors and publisher cannot assume any responsibility for the validity of all materials or the consequences of their use.

**Trademark Notice:** Registered trademark of products or corporate names are used only for explanation and identification without intent to infringe.

**Cataloging-in-Publication Data**

Discrete mathematics / Carlos Davenport.
p. cm.
Includes bibliographical references and index.
ISBN 978-1-63238-878-0
1. Discrete mathematics. 2. Numerical analysis. 3. Mathematics. I. Davenport, Carlos.
QA297.4 .D57 2021
511.1--dc23

# Table of Contents

# Preface

This book is a culmination of my many years of practice in this field. I attribute the success of this book to my support group. I would like to thank my parents who have showered me with unconditional love and support and my peers and professors for their constant guidance.

The study of mathematical structures which are basically discrete instead of continuous is known as discrete mathematics. The objects of study within this field possess distinct, separated values. The sets of these objects can be either finite or infinite. Some of the major topics of study within this field are combinatorics, set theory, theoretical computer science, logic, number theory and probability. Combinatorics is involved in the study of the way in which discrete structures can be arranged or combined. Set theory in discrete mathematics primarily deals with the finite sets. Principles and concepts from this field are applied in numerous areas such as cryptography, computer algorithms, programming languages and software development. This book provides significant information of this discipline to help develop a good understanding of discrete mathematics. It covers in detail some existent theories and innovative concepts revolving around this field. Those in search of information to further their knowledge will be greatly assisted by this book.

The details of chapters are provided below for a progressive learning:

Chapter - What is Discrete Mathematics?

Discrete mathematics is the study of mathematical structures which are discrete in nature. It is used in solving problems of computer algorithms, programming languages, cryptography, etc. The topics elaborated in this chapter will help in gaining a better perspective about discrete mathematics.

Chapter - Set Theory

Set theory is defined as a branch of mathematical theory that deals with the study of collection of objects called sets. There are various types of sets which can be categorized as empty set, power set, transitive set, paradoxical set, etc. This chapter discusses in detail all the aspects related to set theory.

Chapter - Number Theory

Number theory is the branch of pure mathematics that is concerned with the study of integers and integer valued functions. A few of its sub-divisions include lgebraic number theory, analytic number theory, additive number theory, combinatorial number theory, etc. All these sub-divisions of number theory have been carefully analyzed in this chapter.

Chapter - Combinatorics

Combinatorics is the field of mathematics that deals with the problems of selection, configuration and operation within a discrete system. Some of its important aspects include combinatorial modeling, Pascal's identity, permutation, etc. This chapter closely examines these key aspects of combinatorics to provide an extensive understanding of the subject.

Chapter - Graph Theory

The relation between the pairwise objects can be modeled and studied under graph theory. Graph factorization, graph isomorphism, graph automorphism, graph amalgamation, etc. are some of the concepts of graph theory. All these concepts of graph theory and types of graphs have been carefully analyzed in this chapter.

Chapter - Probability

The branch of mathematics that is concerned with the study of possible outcomes of dependent and independent events is known as probability. It includes probability distribution, binomial distribution, Bayes' theorem, conditional probability etc. This chapter has been carefully written to provide an easy understanding of these aspects of probability.

**Carlos Davenport**

# 1

# What is Discrete Mathematics?

Discrete mathematics is the study of mathematical structures which are discrete in nature. It is used in solving problems of computer algorithms, programming languages, cryptography, etc. The topics elaborated in this chapter will help in gaining a better perspective about discrete mathematics.

Discrete mathematics is the study of mathematical structures that are countable or otherwise distinct and separable. Examples of structures that are discrete are combinations, graphs, and logical statements. Discrete structures can be finite or infinite. Discrete mathematics is in contrast to continuous mathematics, which deals with structures which can range in value over the real numbers, or have some non-separable quality.

Since the time of Isaac Newton and until quite recently, almost the entire emphasis of applied mathematics has been on continuously varying processes, modeled by the mathematical continuum and using methods derived from the differential and integral calculus. In contrast, discrete mathematics concerns itself mainly with finite collections of discrete objects. With the growth of digital devices, especially computers, discrete mathematics has become more and more important.

Discrete structures can be counted, arranged, placed into sets, and put into ratios with one another. Although discrete mathematics is a wide and varied field, there are certain rules that carry over into many topics. The concept of independent events and the rules of product, sum, and PIE are shared among combinatorics, set theory, and probability. In addition, De Morgan's laws are applicable in many fields of discrete mathematics.

Often, what makes discrete mathematics problems interesting and challenging are the restrictions that are placed on them. Although the field of discrete mathematics has many elegant formulas to apply, it is rare that a practical problem will fit perfectly to a specific formula. Part of the joy of discovering discrete mathematics is to learn many different approaches to problem-solving, and then be able to creatively apply disparate strategies towards a solution.

## Combinatorics

Combinatorics is the mathematics of counting and arranging. Of course, most people know how to count, but combinatorics applies mathematical operations to count things that are much too large to be counted the conventional way.

Combinatorics is especially useful in computer science. Combinatorics methods can be used to develop estimates about how many operations a computer algorithm will require. Combinatorics is also important for the study of discrete probability. Combinatorics methods can be used to count possible outcomes in a uniform probability experiment.

Combinatorics is often concerned with how things are arranged. In this context, an arrangement is a way objects could be grouped. The most basic rules regarding arrangements are the rule of product and the rule of sum. These rules govern how to count arrangements using the operations of multiplication and addition, respectively.

A more specific type of arrangement is a permutation. A permutation is an arrangement of objects with regard to order.

A combination is another type of arrangement that is related to permutations. A combination is an arrangement of objects without regard to order.

As a field of mathematics, combinatorics is nearly as broad as discrete mathematics. Other topics within combinatorics include:

- Derangements: A permutation such that no object is in its original spot in the order;
- Rectangular grid walks: Determining the number of ways a rectangular lattice can be traversed;
- Distribution of objects into bins: determining how objects can be grouped into bins.

## Set Theory

Set theory is the branch of mathematics that is concerned about collections of objects. Sets can be discrete or continuous; discrete mathematics is primarily concerned with the former. At a basic level, set theory is concerned with how sets can be arranged, combined, and counted.

The cardinality of a finite set is the number of elements in that set. For a given set, $A$, its cardinality is denoted by $|A|$.

Example: What is the cardinality of the set of prime numbers less than 25?

The set of prime numbers less than 25 is:

$$\{2,3,5,7,11,13,17,19,23\}.$$

There are 9 elements in this set, so the cardinality is 9.

Cardinality can also be extended to infinite sets. Although this kind of cardinality cannot be counted, each cardinality can be compared with another cardinality.

Let $A$ and $B$ be sets. Their cardinalities are compared as follows:

If there exists a bijection between $A$ and, $B$, then $|A|=|B|$. If there exists an injective function from $A$ to $B$, but no bijective function, then $|A|<|B|$.

Example: Show that the set of integers and the set of even integers have the same cardinality.

It might seem strange that these sets have the same cardinality. After all, the even integers are more "rare." However, these sets are both *infinite*. Therefore, the "common sense" thinking about *finite* sets must be discarded. Instead, the goal is to obtain a bijective function from the set of integers to the set of even integers:

$$f(n) = 2n, n \in \mathbb{Z}$$

The function above gives a one-to-one correspondence between each integer $n$ and each even integer $2n$. Since the bijection is established, the set of integers and the set of even integers have the same cardinality.

A complement of a set $A$ is the set of elements that are not in $A$. The study of set complements gives a number of efficient methods to calculate cardinalities of finite sets. For example, one can efficiently obtain the cardinality of a set that contains "at least one" element of another set.

The union and intersection give ways to describe how sets can be combined. De Morgan's laws give identities for the complements of unions and intersections.

The principle of inclusion and exclusion, or PIE, gives a method to find the union or intersection of more than two sets.

## Graph Theory

Graph theory is the study of graphs, which are a collection of connected nodes.

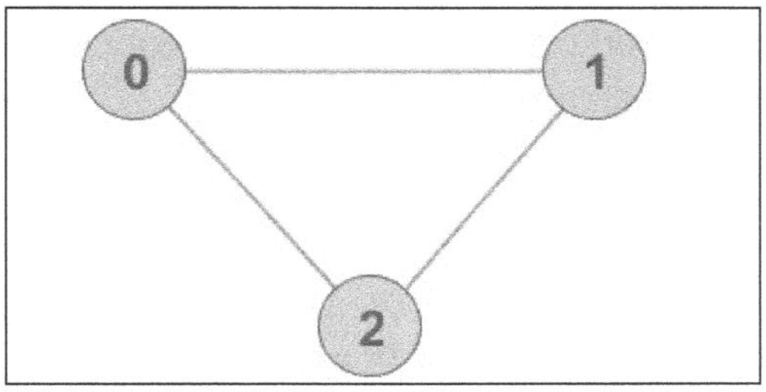


A basic graph of 3-cycle.

Graphs are useful for representing all kinds of real-world problems.

## Probability

A probability is a number, between 0 and 1 inclusive, that represents the likelihood of an event. Discrete probability is a probability based on discrete sets of outcomes. The most basic type of probability is a uniform probability. If each outcome in a set is equally likely, then the probability of an event is equal to a ratio of cardinalities.

Let $S$ be a sample space of outcomes. If each outcome in this set is equally likely, then the probability of an event $A$ in $S$ is:

$$P(A) = \frac{|A|}{|S|}.$$

Many of the rules of probability are analogous to the rules of combinatorics. The probabilistic rules of product, sum, and complement work similarly to those same rules from combinatorics. In addition, the structure of the probabilistic principle of inclusion and exclusion is the same as PIE for sets.

A discrete probability distribution is a function that takes a numerical outcome as an argument and gives a probability as a result. Discrete probability distributions can be created using the rules and guidelines. There are also some discrete probability distributions that show up in many problems:

- Geometric Distribution: Given repeated trials in which the probability of success is the same each time, this gives the probability that the first success will occur on a certain trial. Example: You roll a dice until you roll a 6. What is the probability that the first 6 will occur on the third roll?
- Binomial Distribution: Given a certain number of trials in which the probability of success is the same each time, this gives the probability of a certain number of successes. Example: You flip a coin 10 times. What is the probability that there will be exactly 5 heads?
- Poisson Distribution: Given a time period in which an event occurs a certain average number of times, this gives the probability that the event will occur a specific number of times. Example: A fast food drive-through gets 3 customers per minute. What's the probability they will get 4 customers in the next minute?

Although basic probabilities are based on discrete sets, the concept of probability can be extended to continuous sets by using concepts from calculus.

## Statistics

A statistic is a number used to describe a set of data or a probability distribution. Statistics is widely used in many fields outside of mathematics, from biology to politics to sports. The power of statistics lies in taking a massive, varied set of data and making sense out of it. Furthermore, statistics has the power to quantify *confidence* in those findings. Of course, the usefulness of statistics is not without controversy, but an understanding of its theoretical underpinnings can help one avoid its misuse.

One major kind of statistic is a measure of central tendency. A measure of central tendency is a number which describes what a value of a probability distribution or data set will tend to. An expected value is the theoretical long-run average outcome of a probability experiment when it is performed many times.

Somewhat related to the expected value is the mean. The mean is the average value of a set of numerical data.

Another major kind of statistic is a measure of variation. A measure of variation is a number which describes the distribution of a probability distribution or data set. The standard deviation of a probability distribution is a number that represents how much the outcomes differ from the expected value. Likewise, the standard deviation of a data set is a number that represents how much the elements of the set differ from the mean.

Although discrete statistics are based off of discrete events and probability distributions, these same concepts can be extended to continuous events and probability distributions using concepts from calculus.

## Bijections

A bijection is a relationship between two sets such that each element in a set is paired with exactly one element in the other set, and vice versa. Bijections can be applied to problem solving by establishing a bijection between a set that is difficult to enumerate and a discrete stucture that is well understood. By establishing a bijection, one can take advantage of the known formulas and theorems that the discrete structure affords.

The three Molloy siblings, April, Bradley, and Clark, have integer ages that sum to 15. How many possible distribution of ages are there?

One can establish a bijection between the set of distributions of ages and a set of combinations. Consider the arrangement of stars and bars below:

**|****|*********

This arrangement corresponds to the following distribution of ages: April - 2, Bradley - 4, Clark - 9. Note that there are 15 stars and 2 bars in the arrangement above. This gives a total of 17 objects, 2 of which are bars. Placing the bars in different spots among the 17 placements will give a new distribution of ages. Thus, a bijection can be established between the set of distributions of ages and the set of combinations of 2 objects out of 17.

The number of distributions of ages is:

$$\binom{17}{2} = 136.$$

## Logic

A proposition is a statement that can either be true or false. Propositional logic aims to outline the rules of how these statements can be altered and combined.

Similarly, Boolean algebra outlines the operations defined on variables that can take the values of true (1) or false (0). Boolean algebra is used to design computer circuits through logic gates, which take signal(s) as inputs and return a signal as an output.

## Functions

It is a mapping in which every element of set A is uniquely associated at the element with set B. The set of A is called Domain of a function and set of B is called Co domain.

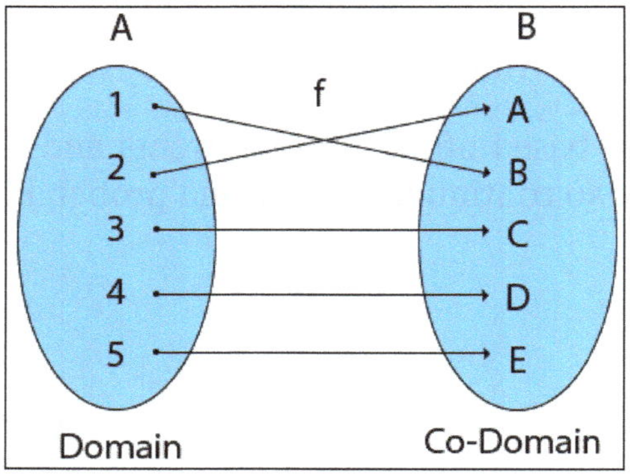


## Domain, Co-domain and Range of a Function

Domain of a Function: Let f be a function from P to Q. The set P is called the domain of the function f.

Co-domain of a Function: Let f be a function from P to Q. The set Q is called Co-domain of the function f.

Range of a Function: The range of a function is the set of picture of its domain. In other words, we can say it is a subset of its co-domain. It is denoted as f (domain).

If f: $P \to Q$, then $f(P) = \{f(x): x \in P\} = \{y: y \in Q \mid \exists\, x \in P, \text{ such that } f(x) = y\}$.

Example: Find the Domain, Co-domain, and Range of function.

Let, x = {1, 2, 3, 4}

y = {a, b, c, d, e}

f = {(1, b), (2, a), (3, d), (4, c)}

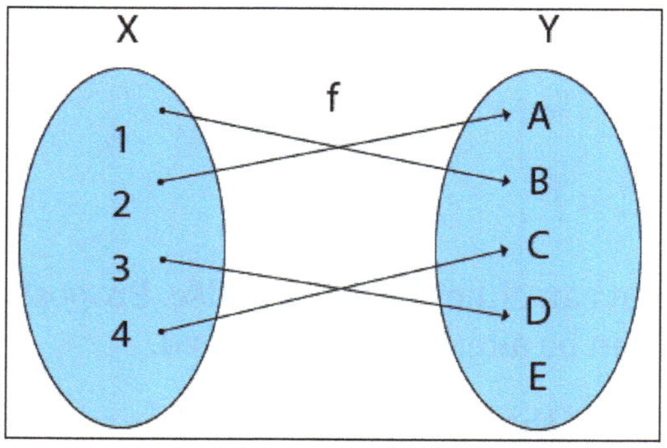


Solution:

Domain of function: {1, 2, 3, 4}

Range of function: {a, b, c, d}

Co-domain of function: {a, b, c, d, e}

## Functions as a Set

If P and Q are two non-empty sets, then a function f from P to Q is a subset of P x Q, with two important restrictions:

- $\forall a \in P$, $(a, b) \in f$ for some $b \in Q$.
- If $(a, b) \in f$ and $(a, c) \in f$ then $b = c$.

Note:

- There may be some elements of the Q which are not related to any element of set P.
- Every element of P must be related with at least one element of Q.

Example: If a set A has n elements, how many functions are there from A to A?

Solution: If a set A has n elements, then there are $n^n$ functions from A to A.

## Representation of a Function

The two sets P and Q are represented by two circles. The function $f: P \to Q$ is represented by a collection of arrows joining the points which represent the elements of P and corresponds elements of Q.

Example:

Let,

$X = \{a, b, c\}$ and $Y = \{x, y, z\}$ and $f: X \to Y$,

such that,

$f= \{(a, x), (b, z), (c, x)\}$

Then f can be represented diagrammatically as follows:

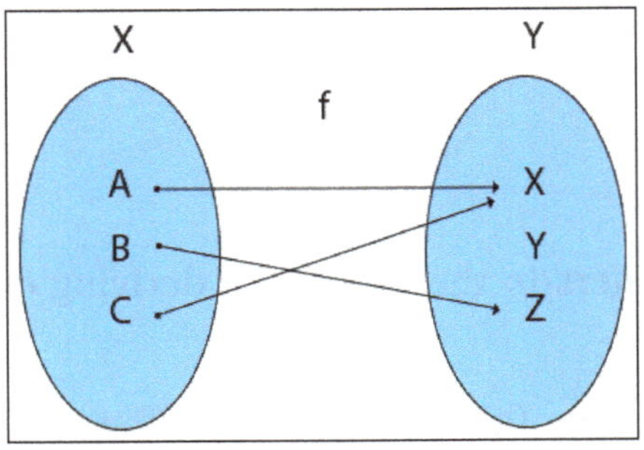


Example: Let $X = \{x, y, z, k\}$ and $Y = \{1, 2, 3, 4\}$. Determine which of the following functions. Give reasons if it is not. Find range if it is a function.

- $f = \{(x, 1), (y, 2), (z, 3), (k, 4)\}$
- $g = \{(x, 1), (y, 1), (k, 4)\}$
- $h = \{(x, 1), (x, 2), (x, 3), (x, 4)\}$

- l = {(x, 1), (y, 1), (z, 1), (k, 1)}
- d = {(x, 1), (y, 2), (y, 3), (z, 4), (z, 4)}

Solution:

- It is a function. Range (f) = {1, 2, 3, 4}.
- It is not a function because every element of X does not relate with some element of Y i.e., Z is not related with any element of Y.
- h is not a function because h (x) = {1, 2, 3, 4} i.e., element x has more than one image in set Y.
- d is not a function because d (y) = {2, 3} i.e., element y has more than image in set Y.

## Algorithms and Functions

An algorithm is a step-by-step method for solving some problem.

Algorithms generally have the following characteristics:

- Input: The algorithm receives input. Zero or more quantities are externally supplied.
- Output: The algorithm produces output. At least one quantity is produced.
- Precision: The steps are precisely stated. Each instruction is clear and unambiguous.
- Feasibility: It must be feasible to execute each instruction.
- Flexibility: It should also be possible to make changes in the algorithm without putting so much effort on it.
- Generality: The algorithm applies to a set of inputs.
- Finiteness: Algorithm must complete after a finite number of instruction have been executed.

## Analysis of Algorithms

The Analysis of an algorithm refers to the process of deriving estimates for the time and space needed to execute the algorithm.

It is important to estimate the time (e.g., the number of steps) and space (e.g., the number of variables) required by algorithms. Knowing the time and space required by algorithm allows us to compare the algorithms that solve the same problem. For example, if one algorithm takes n steps to solve a problem and another algorithm takes $n^2$ steps to solve the same problem, we would prefer the first algorithm. This estimation of time and space needed to execute the algorithm is called the time and space complexity of the algorithm.

The time required to execute an algorithm is a function of the input. Instead of dealing directly with the input, parameters are used to characterize the size of the input. e.g. if the input is a set containing n elements, the size of the input n. There are three cases worth noting about the time complexity of an algorithm since determining the exact time complexity of an algorithm in a difficult task.

- Worst-case: f (n) represent by the maximum number of steps taken on any instance of size n.
- Best-case: f (n) represent by the minimum number of steps taken on any instance of size n.
- Average case: f (n) represent by the average number of steps taken on any instance of size n.

## Asymptotic Notations

Asymptotic Notations are used to describe the execution time of an algorithm. The notations show the order of growth of functions. Here the time taken by an algorithm is mapped regarding mathematical functions. There are many asymptotic notations like o, θ, Ω, w each having its importance.

1. Big-oh notation: The function f (n) =O (g (n)), if and only if exist positive constant c and $n_0$ such that:

$$f(n) \leqslant C \times g(n) \text{ for all } n, n \geq n_0$$

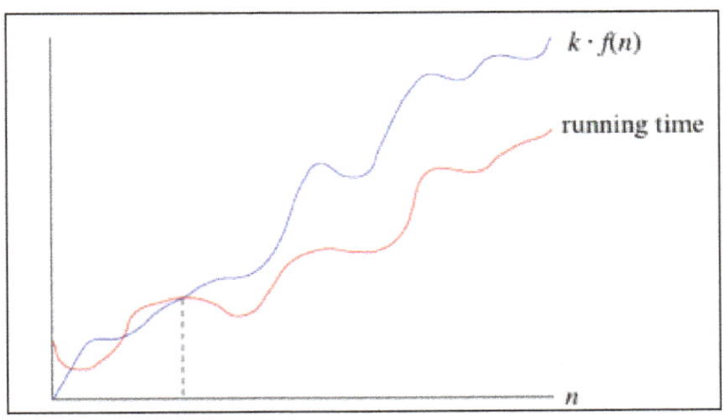


Asymptotic upper bound.

Example: The function 4n + 3 = O (n) as $4n + 3 \leqslant 5n$ for all n ≥.

Example: The function $20n^2 + 5n + 2 = O(n^2)$ as $20n^2 + 5n + 2 \leqslant 21n^2$ for all n ≥.

2. Omega (Ω) Notation: The function f (n) = Ω (g (n)) [read as "f of n is omega of g of n"] if and only if there exists positive constant c and $n_0$ such that:

$$f(n) \geq C^* g(n) \text{ for all } n, n \geq n_0$$

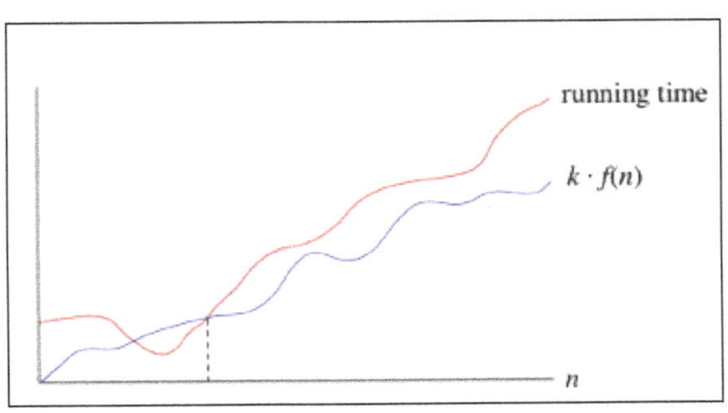


Asymptotic lower bound.

Example: The function $4n + 3 = \Omega(n)$ as $4n + 3 \geq 4n$ for all $n \geq 1$.

Example: The function $20n^2 + 5n + 2 = \Omega(n)$ as $20n^2 + 5n + 2 \geq 20n^2$ for all $n \geq 1$.

3. Theta ($\theta$): The function $f(n) = \theta(g(n))$, if and only if there exists positive constant $k_1$, $k_2$ and $k_0$ such that:

$$k_1 * g(n) \leq f(n) \leq k_2 * g(n) \text{ for all } n, n \geq n_0$$

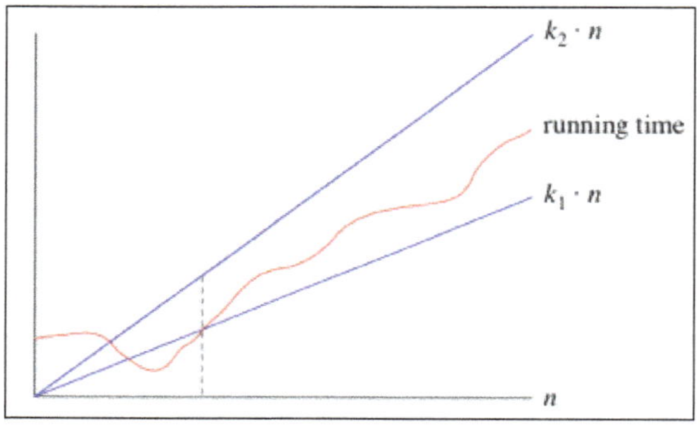


Asymptotic tight bound.

Example: The function $4n + 3 = \theta(n)$ as $4n + 3 \geq 4n$ for all $n \geq 3$ and $4n + 3 \leq 5n$ for all $n \geq 3$.

Example: The function $20n^2 + 5n + 2 = \theta(n^2)$ as $20n^2 + 5n + 2 \leq 21n^2$ for all $n \geq 1$ and $20n^2 + 5n + 2 \geq 20n^2$ for all $n \geq 1$.

## Basic Counting Principles

Sum Rule Principle: Assume some event E can occur in m ways and a second event F can occur in n ways, and suppose both events cannot occur simultaneously. Then E or F can occur in m + n ways.

In general, if there are n events and no two events occurs in same time then the event can occur in $n_1 + n_2 .........n$ ways.

Example: If 8 male processor and 5 female processor teaching DMS then the student can choose professor in 8 + 5 = 13 ways.

Product Rule Principle: Suppose there is an event E which can occur in m ways and, independent of this event, there is a second event F which can occur in n ways. Then combinations of E and F can occur in mn ways.

In general, if there are n events occurring independently then all events can occur in the order indicated as $n_1 \times n_2 \times n_3 .........n$ ways.

Example: In class, there are 4 boys and 10 girls if a boy and a girl have to be chosen for the class monitor, the students can choose class monitor in 4 x 10 = 40 ways.

## Mathematical Functions

Factorial Function: The product of the first n natural number is called factorial n. It is denoted by n!, read "n Factorial."

The Factorial n can also be written as:

$$n! = n\,(n-1)\,(n-2)\,(n-3)......1.$$

$$= 1 \text{ and } 0! = 1.$$

Example: Find the value of 5!

Solution:

$$5! = 5 \times (5-1)\,(5-2)\,(5-3)\,(5-4)$$

$$= 5 \times 4 \times 3 \times 2 \times 1 = 120$$

Example: Find the value of $\frac{10!}{8!}$

Solution: $\frac{10!}{8!} = \frac{10 \times 9 \times 8}{8!} = 10 \times 9 = 90$

Binomial Coefficients: Binomial Coefficient is represented by $n_{C_r}$ where r and n are positive integer with $r \leq n$ is defined as follows:

$$\binom{n}{r} = \frac{n(n-1)(n-2)............(n-r+1)}{1*2*3............(r-1)r}$$

Example: $8_{C_2} = \frac{8!}{2!6!} = \frac{8 \times 7 \times 6}{2 \times 6} = 28.$

# 2
# Set Theory

Set theory is defined as a branch of mathematical theory that deals with the study of collection of objects called sets. There are various types of sets which can be categorized as empty set, power set, transitive set, paradoxical set, etc. This chapter discusses in detail all the aspects related to set theory.

Any well-defined collection of mathematical objects can form a set. These objects could be anything – from people's names to their ages/likes /dislikes; entities from simple number system to complex scientific data; from outcomes of a single dice roll or a coin toss to such experiments repeated 100s or 1000s of times.

The only condition which is to be kept in mind is that the entities or objects must be related through the same rule. For example:

- Family of all natural numbers/whole numbers/odd numbers/even numbers/rational numbers/integers/real numbers.
- A group of possible outcomes of a dice roll or a coin toss.
- Collection of crucial data gathered by ISRO from MOM.
- A collection of day/night temperatures.

And there are many more such examples that will form a collection. The important point to notice here is that the rule that defines or law through which we collect or group objects should be universal.

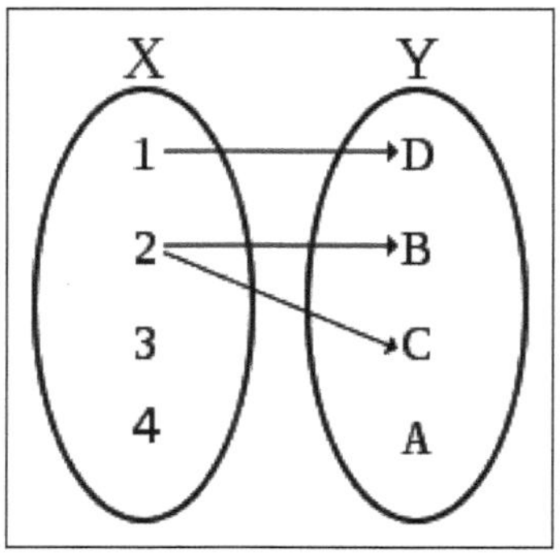

## Conventions for Sets

The following are the conventions that are used here:

- Sets are *usually* denoted by a capital letter.
- The elements of the group are *usually* represented by small letters (unless specified separately).
- If 'a' is an element of 'A', or if a "belongs to" A, it is written in the conventional notion by the use of the symbol ϵ (Epsilon) between them – a ϵ A.
- If b is not an element of Set A, b "does not belong to" A is written in the conventional notion by the use of the symbol ϵ (Epsilon with a line across it) between them – a ϵ A.
- Objects, elements, entities, members are all synonymous terms.

## Representations of a Set

Representation of Sets and its elements is done in the following two ways.

Roster Form: In this form, all the elements are enclosed within braces {} and they are separated by commas (,). For example, a collection of all the numbers found on a dice N = {1, 2, 3, 4, 5, 6}. Properties of roster form:

- The order in which the elements are listed in the Roster form for any Set is immaterial. For example, V = {a, e, i, o, u} is same as V = {u, o, e, a, i}.
- The dots at the end of the last element of any Set represent its infinite form and indefinite nature. For example, group of odd natural numbers = {1, 3, 5, ...}.
- In this form of representation, the elements are *generally* not repeated. For example, the group of letters forming the word POOL = {P, O, L}.

More examples for Roster form of representation are:

- A = {3, 6, 9, 12}
- F = {2, 4, 8, 16, 32}
- H = {1, 4, 9, 16, ..., 100}
- L = {5, 25, 125, 625}
- Y = {1, 1, 2, 3, 5, 8, ...}

Set Builder Form: In this form, all the elements possess a single common property which is not featured by any other element outside the Set. For example, a group of vowels in English alphabetical series.

The representation is done as follows. Let V be the collection of all English vowels, then – V = {x: x is a vowel in English alphabetical series.} Properties of Roster form:

- Colon (:) is a mandatory symbol for this type of representation.
- After the colon sign, we write the *common characteristic property* possessed by all the elements belonging to that Set and enclose it within braces.
- If the Set doesn't follow a pattern, its Set builder form cannot be written.

More examples for Set builder form of representation for a Set:

- D = {x: x is an integer and – 3 < x < 19}.
- O = {y: y is a natural number greater than 5}.
- I = {f: f is a two – digit prime number less than 1000}.
- R = {s: s is a natural number such that sum of its digits is 4}.
- X = {m: m is a positive integer < 40}.

## Set Theory

Set theory is a branch of mathematical logic that studies sets, which informally are collections of objects. Although any type of object can be collected into a set, set theory is applied most often to objects that are relevant to mathematics. The language of set theory can be used to define nearly all mathematical objects.

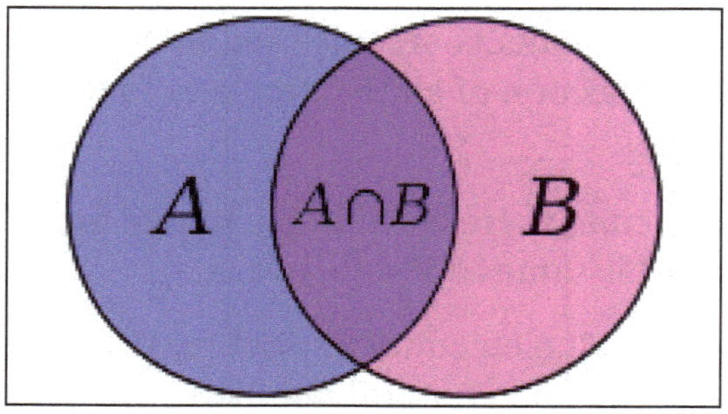


A Venn diagram illustrating the intersection of two sets.

The modern study of set theory was initiated by Georg Cantor and Richard Dedekind in the 1870s. After the discovery of paradoxes in naive set theory, such as Russell's paradox, numerous axiom systems were proposed in the early twentieth century, of which the Zermelo–Fraenkel axioms, with or without the axiom of choice, are the best-known.

Set theory is commonly employed as a foundational system for mathematics, particularly in the form of Zermelo–Fraenkel set theory with the axiom of choice. Beyond its foundational role, set theory is a branch of mathematics in its own right, with an active research community. Contemporary research into set theory includes a diverse collection of topics, ranging from the structure of the real number line to the study of the consistency of large cardinals.

### Axiomatic Set Theory

Elementary set theory can be studied informally and intuitively, and so can be taught in primary schools using Venn diagrams. The intuitive approach tacitly assumes that a set may be formed from the class of all objects satisfying any particular defining condition. This assumption gives rise to paradoxes, the simplest and best known of which are Russell's paradox and the Burali-Forti paradox. Axiomatic set theory was originally devised to rid set theory of such paradoxes.

The most widely studied systems of axiomatic set theory imply that all sets form a cumulative hierarchy. Such systems come in two flavors, those whose ontology consists of:

- Sets alone: This includes the most common axiomatic set theory, Zermelo–Fraenkel set theory (ZFC), which includes the axiom of choice. Fragments of ZFC include:
    - Zermelo set theory, which replaces the axiom schema of replacement with that of separation;
    - General set theory, a small fragment of Zermelo set theory sufficient for the Peano axioms and finite sets;
    - Kripke–Platek set theory, which omits the axioms of infinity, powerset, and choice, and weakens the axiom schemata of separation and replacement.
- Sets and proper classes: These include Von Neumann–Bernays–Gödel set theory, which has the same strength as ZFC for theorems about sets alone, and Morse–Kelley set theory and Tarski–Grothendieck set theory, both of which are stronger than ZFC.

The above systems can be modified to allow urelements, objects that can be members of sets but that are not themselves sets and do not have any members.

The systems of New Foundations NFU (allowing urelements) and NF (lacking them) are not based on a cumulative hierarchy. NF and NFU include a "set of everything," relative to which every set has a complement. In these systems urelements matter, because NF, but not NFU, produces sets for which the axiom of choice does not hold.

Systems of constructive set theory, such as CST, CZF, and IZF, embed their set axioms in intuitionistic instead of classical logic. Yet other systems accept classical logic but feature a nonstandard membership relation. These include rough set theory and fuzzy set theory, in which the value of an atomic formula embodying the membership relation is not simply true or false. The Boolean-valued models of ZFC are a related subject. An enrichment of ZFC called internal set theory was proposed by Edward Nelson in 1977.

## Applications

Many mathematical concepts can be defined precisely using only set theoretic concepts. For example, mathematical structures as diverse as graphs, manifolds, rings, and vector spaces can all be defined as sets satisfying various (axiomatic) properties. Equivalence and order relations are ubiquitous in mathematics, and the theory of mathematical relations can be described in set theory.

Set theory is also a promising foundational system for much of mathematics. Since the publication of the first volume of *Principia Mathematica*, it has been claimed that most or even all mathematical theorems can be derived using an aptly designed set of axioms for set theory, augmented with many definitions, using first or second-order logic. For example, properties of the natural and real numbers can be derived within set theory, as each number system can be identified with a set of equivalence classes under a suitable equivalence relation whose field is some infinite set.

Set theory as a foundation for mathematical analysis, topology, abstract algebra, and discrete mathematics is likewise uncontroversial; mathematicians accept that (in principle) theorems in these areas can be derived from the relevant definitions and the axioms of set theory. Few full derivations of complex mathematical theorems from set theory have been formally verified, however, because such formal derivations are often much longer than the natural language proofs mathematicians commonly present.

## Transitive Set

In set theory, a set $A$ is called transitive if either of the following equivalent conditions hold:

- Whenever $x \in A$, and $y \in x$, then $y \in A$.
- Whenever $x \in A$, and $x$ is not an urelement, then $x$ is a subset of $A$.

Similarly, a class $M$ is transitive if every element of $M$ is a subset of $M$.

Examples: Using the definition of ordinal numbers suggested by John von Neumann, ordinal numbers are defined as hereditarily transitive sets: an ordinal number is a transitive set whose members are also transitive (and thus ordinals). The class of all ordinals is a transitive class.

Any of the stages $V_\alpha$ and $L_\alpha$ leading to the construction of the von Neumann universe $V$ and Gödel's constructible universe $L$ are transitive sets. The universes $L$ and $V$ themselves are transitive classes.

This is a complete list of all finite transitive sets with up to 20 brackets:

- {},
- {{}},
- {{},{{}}},
- {{},{{}},{{{}}}},
- {{},{{}},{{},{{}}}},
- {{},{{}},{{{}}},{{{{}}}}},
- {{},{{}},{{{}}},{{},{{}}}},
- {{},{{}},{{{}}},{{},{{{}}}}},
- {{},{{}},{{{}}},{{{}},{{{}}}}},
- {{},{{}},{{{},{{}}}},{{},{{}}}},
- {{},{{}},{{{}}},{{},{{}},{{{}}}}},
- {{},{{}},{{},{{}}},{{},{{},{{}}}}},

- {{},{{}},{{},{{}}},{{{}},{{},{{}}}}},
- {{},{{}},{{{}}},{{{{}}}},{{},{{}}}},
- {{},{{}},{{},{{}}},{{},{{}},{{},{{}}}}},
- {{},{{}},{{{}}},{{{{}}}},{{{{{}}}}}},
- {{},{{}},{{{}}},{{{{}}}},{{},{{{}}}}},
- {{},{{}},{{{}}},{{{},{{}}}},{{},{{}}}},
- {{},{{}},{{{}}},{{},{{}}},{{},{{{}}}}},
- {{},{{}},{{{}}},{{{{}}}},{{},{{{{}}}}}},
- {{},{{}},{{{}}},{{{{}}}},{{{}},{{{}}}}},
- {{},{{}},{{{}}},{{},{{}}},{{},{{},{{}}}}},
- {{},{{}},{{{}}},{{},{{}}},{{{}},{{{}}}}},
- {{},{{}},{{{}}},{{{{}}}},{{{}},{{{{}}}}}},
- {{},{{}},{{{}}},{{{{}}}},{{},{{}},{{{}}}}},
- {{},{{}},{{{}}},{{{},{{{}}}}},{{},{{{}}}}},
- {{},{{}},{{{}}},{{},{{}}},{{{}},{{},{{}}}}},
- {{},{{}},{{{}}},{{},{{}}},{{},{{}},{{{}}}}},
- {{},{{}},{{{}}},{{},{{{}}}},{{{}},{{{}}}}},
- {{},{{}},{{{}}},{{{{}}}},{{{{}}},{{{{}}}}}},
- {{},{{}},{{{}}},{{{{}}}},{{},{{}},{{{{}}}}}},
- {{},{{}},{{{}}},{{},{{}}},{{{{}}},{{},{{}}}}},
- {{},{{}},{{{}}},{{},{{}}},{{},{{}},{{},{{}}}}},
- {{},{{}},{{{}}},{{},{{{}}}},{{},{{},{{{}}}}}},
- {{},{{}},{{{}}},{{},{{{}}}},{{},{{}},{{{}}}}},
- {{},{{}},{{{{},{{}}}}},{{{},{{}}}},{{},{{}}}},
- {{},{{}},{{{},{{}}}},{{},{{}}},{{},{{},{{}}}}},
- {{},{{}},{{{}}},{{{{}}}},{{},{{{}}},{{{{}}}}}},
- {{},{{}},{{{}}},{{{{}},{{{}}}}},{{{}},{{{}}}}},
- {{},{{}},{{{}}},{{},{{}}},{{},{{{}}},{{},{{}}}}},

- $\{\{\},\{\{\}\},\{\{\{\}\}\},\{\{\},\{\{\{\}\}\}\},\{\{\{\}\},\{\{\},\{\{\{\}\}\}\}\}\}$,
- $\{\{\},\{\{\}\},\{\{\{\}\}\},\{\{\{\}\},\{\{\{\}\}\}\},\{\{\},\{\{\}\},\{\{\{\}\}\}\}\}$,
- $\{\{\},\{\{\}\},\{\{\{\},\{\{\}\}\}\},\{\{\},\{\{\}\}\},\{\{\},\{\{\{\},\{\{\}\}\}\}\}\}$,
- $\{\{\},\{\{\}\},\{\{\{\}\}\},\{\{\{\{\}\}\}\},\{\{\{\{\{\}\}\}\}\},\{\{\},\{\{\}\}\}\}$,
- $\{\{\},\{\{\}\},\{\{\{\}\}\},\{\{\{\{\}\}\}\},\{\{\{\},\{\{\}\}\}\},\{\{\},\{\{\}\}\}\}$,
- $\{\{\},\{\{\}\},\{\{\{\}\}\},\{\{\{\{\}\}\}\},\{\{\},\{\{\}\}\},\{\{\},\{\{\{\}\}\}\}\}$.

## Properties

A set $X$ is transitive if and only if $\bigcup X \subseteq X$, where $\bigcup X$ is the union of all elements of $X$ that are sets, $\bigcup X = \{y \mid \exists x \in X : y \in x\}$. If $X$ is transitive, then $\bigcup X$ is transitive. If $X$ and $Y$ are transitive, then $X \cup Y \cup \{X,Y\}$ is transitive. In general, if $X$ is a class all of whose elements are transitive sets, then $X \cup \bigcup X$ is transitive.

A set $X$ which does not contain urelements is transitive if and only if it is a subset of its own power set, $X \subset \mathcal{P}(X)$. The power set of a transitive set without urelements is transitive.

## Transitive Closure

The transitive closure of a set $X$ is the smallest (with respect to inclusion) transitive set which contains $X$. Suppose one is given a set $X$, then the transitive closure of $X$ is:

$$\mathrm{TC}(X) = \bigcup\left\{X, \bigcup X, \bigcup\bigcup X, \bigcup\bigcup\bigcup X, \bigcup\bigcup\bigcup\bigcup X, \ldots\right\}.$$

Denote $X_0 = X$ and $X_{n+1} = \bigcup X_n$. Then we claim that the set,

$$T = \mathrm{TC}(X) = \bigcup_{n=0}^{\infty} X_n$$

is transitive, and whenever $T_1$ is a transitive set containing $X$ then $T \subseteq T_1$.

Assume $y \in x \in T$. Then $x \in X_n$ for some $n$ and so $y \in \bigcup X_n = X_{n+1}$. Since $X_{n+1} \subseteq T$, $y \in T$. Thus $T$ is transitive.

Now let $T_1$ be as above. We prove by induction that $X_n \subseteq T_1\ \forall n$ thus proving that $T \subseteq T_1$. Since $X_0 = X \subseteq T_1$ we have a base. Now assume $X_n \subseteq T_1$. Then $X_{n+1} = \bigcup X_n \subseteq \bigcup T_1$. But $T_1$ is transitive so $T_1 \subseteq T_1$ hence $X_{n+1} \subseteq T_1$.

Note that this is the set of all of the objects related to $X$ by the transitive closure of the membership relation.

## Transitive Models of Set Theory

Transitive classes are often used for construction of interpretations of set theory in itself, usually called inner models. The reason is that properties defined by bounded formulas are absolute for transitive classes.

A transitive set (or class) that is a model of a formal system of set theory is called a transitive model of the system (provided that the element relation of the model is the restriction of the true element relation to the universe of the model). Transitivity is an important factor in determining the absoluteness of formulas.

In the superstructure approach to non-standard analysis, the non-standard universes satisfy strong transitivity.

# Paradoxical Set

In set theory, a paradoxical set is a set that has a paradoxical decomposition. A paradoxical decomposition of a set is two families of disjoint subsets, along with appropriate group actions that act on some universe (of which the set in question is a subset), such that each partition can be mapped back onto the entire set using only finitely many distinct functions (or compositions thereof) to accomplish the mapping. A set that admits such a paradoxical decomposition where the actions belong to a group $G$ is called $G$-paradoxical or paradoxical with respect to $G$.

The Banach–Tarski paradox is that a ball can be decomposed into a finite number of point sets and reassembled into two balls identical to the original.

Paradoxical sets exist as a consequence of the Axiom of Infinity. Admitting infinite classes as sets is sufficient to allow paradoxical sets.

Suppose a group acts on a set $A$. Then $A$ is $G$-paradoxical if there exists some disjoint subsets $A_1,\ldots,A_n,B_1,\ldots,B_m \subseteq A$ and some group elements $g_1,\ldots,g_n,h_1,\ldots,h_m \in G$ such that:

$$A=\bigcup_{i=1}^{n} g_i(A_i) \text{ and } A=\bigcup_{i=1}^{m} h_i(B_i)$$

### Free Group

The Free group $F$ on two generators $a$, $b$ has the decomposition $F=\{e\}\cup X(a)\cup X(a^{-1})$ $\cup X(b)\cup X(b^{-1})$ where $e$ is the identity word and $X(i)$ is the collection of all (reduced) words that start with the letter $i$.

This is a paradoxical decomposition because $X(a)\cup aX(a^{-1})=F=X(b)\cup bX(b^{-1})$.

### Banach–Tarski Paradox

The most famous, and indeed motivational, example of paradoxical sets is the Banach–Tarski paradox, which divides the sphere into paradoxical sets for the special orthogonal group. This result depends on the axiom of choice.

## Subset

In mathematics, a set *A* is a subset of a set *B*, or equivalently *B* is a superset of *A*, if *A* is contained in *B*. That is, all elements of *A* are also elements of *B* (note that *A* and *B* may be equal). The relationship of one set being a subset of another is called inclusion or sometimes containment. *A* is a subset of *B* may also be expressed as *B* includes *A*, or *A* is included in *B*.

The subset relation defines a partial order on sets. In fact, the subsets of a given set form a Boolean algebra under the subset relation, in which the meet and join are given by intersection and union.

If *A* and *B* are sets and every element of *A* is also an element of *B*, then:

- *A* is a subset of *B*, denoted by $A \subseteq B$,

or equivalently,

- *B* is a superset of *A*, denoted by $B \supseteq A$.

If *A* is a subset of *B*, but *A* is not equal to *B* (i.e. there exists at least one element of B which is not an element of *A*), then:

- *A* is also a proper (or strict) subset of *B*; this is written as $A \subsetneq B$.

or equivalently,

- *B* is a proper superset of *A*; this is written as $B \supsetneq A$.

For any set *S*, the inclusion relation ⊆ is a partial order on the set $\mathcal{P}(S)$ of all subsets of *S* (the power set of *S*) defined by $A \leq B \Leftrightarrow A \subseteq B$. We may also partially order $\mathcal{P}(S)$ by reverse set inclusion by defining $A \leq B \Leftrightarrow B \subseteq A$.

When quantified, A ⊆ B is represented as: ∀x{x∈A → x∈B}.

### Properties

- A set *A* is a subset of *B* if and only if their intersection is equal to A.

  Formally:

  $$A \subseteq B \Leftrightarrow A \cap B = A.$$

- A set *A* is a subset of *B* if and only if their union is equal to B.

Formally:

$$A \subseteq B \Leftrightarrow A \cup B = B.$$

- A finite set $A$ is a subset of $B$ if and only if the cardinality of their intersection is equal to the cardinality of A.

  Formally:

  $$A \subseteq B \Leftrightarrow |A \cap B| = |A|.$$

## Examples

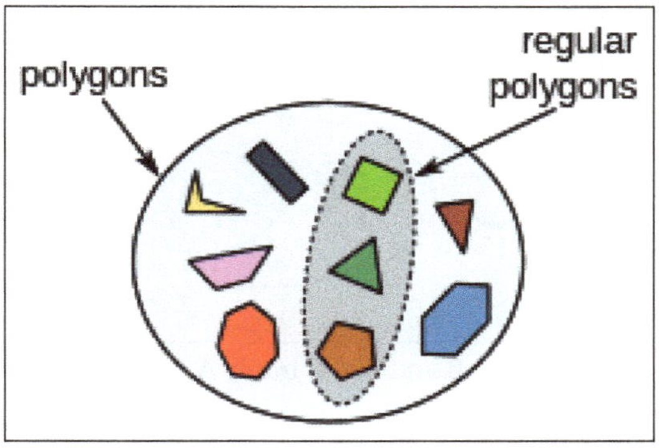


The regular polygons form a subset of the polygons:

- The set A = {1, 2} is a proper subset of B = {1, 2, 3}, thus both expressions $A \subseteq B$ and $A \subsetneq B$ are true.
- The set D = {1, 2, 3} is a subset (but *not* a proper subset) of E = {1, 2, 3}, thus $D \subseteq E$ is true, and $D \subsetneq E$ is not true (false).
- Any set is a subset of itself, but not a proper subset. ($X \subseteq X$ is true, and $X \subsetneq X$ is false for any set X.)
- The empty set { }, denoted by ∅, is also a subset of any given set *X*. It is also always a proper subset of any set except itself.
- The set {*x*: *x* is a prime number greater than 10} is a proper subset of {*x*: *x* is an odd number greater than 10}
- The set of natural numbers is a proper subset of the set of rational numbers; likewise, the set of points in a line segment is a proper subset of the set of points in a line. These are two examples in which both the subset and the whole set are infinite, and the subset has the same cardinality (the concept that corresponds to size, that is, the number of elements, of a finite set) as the whole; such cases can run counter to one's initial intuition.
- The set of rational numbers is a proper subset of the set of real numbers. In this example, both sets are infinite but the latter set has a larger cardinality (or *power*) than the former set.

Another example in an Euler diagram:

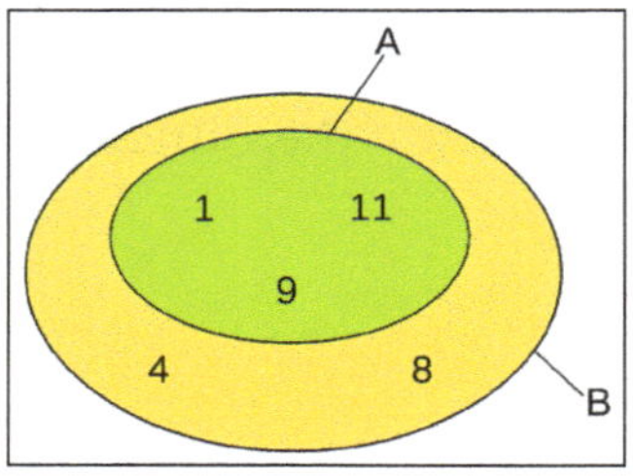


A is a proper subset of B.

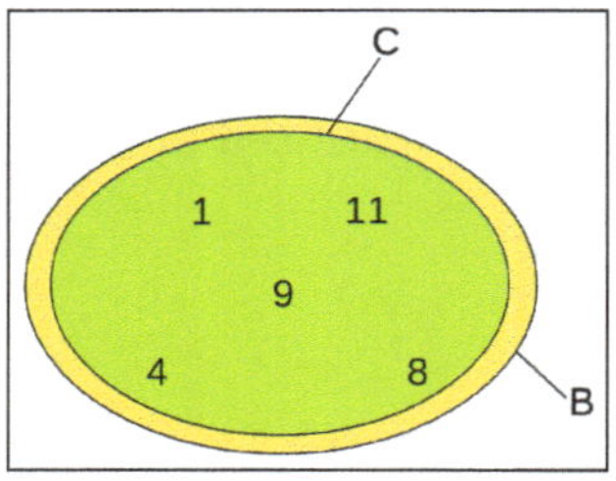


C is a subset but not a proper subset of B.

## Other Properties of Inclusion

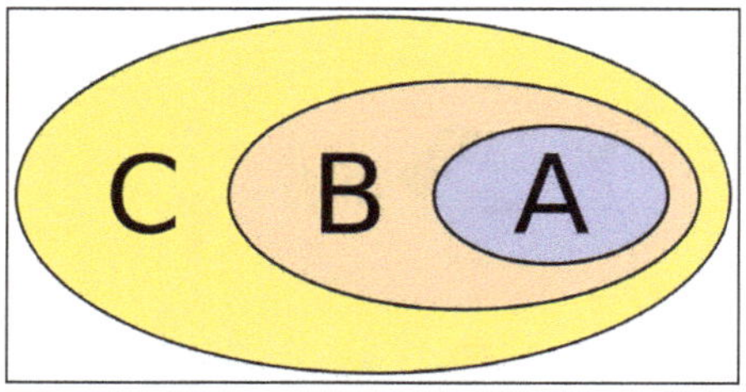


A ⊆ B and B ⊆ C imply A ⊆ C.

Inclusion is the canonical partial order in the sense that every partially ordered set $(X, \prec)$ is isomorphic to some collection of sets ordered by inclusion. The ordinal numbers are a simple example—if each ordinal $n$ is identified with the set $[n]$ of all ordinals less than or equal to $n$, then $a \leq b$ if and only if $[a] \subseteq [b]$.

For the power set $\mathcal{P}(S)$ of a set $S$, the inclusion partial order is (up to an order isomorphism) the Cartesian product of $k = |S|$ (the cardinality of $S$) copies of the partial order on $\{0,1\}$ for which $0 < 1$. This can be illustrated by enumerating $S = \{s_1, s_2, ..., s_k\}$ and associating with each subset $T \subseteq S$ (which is to say with each element of $2^S$) the $k$-tuple from $\{0,1\}^k$ of which the $i$th coordinate is 1 if and only if $s_i$ is a member of $T$.

## Club Set

In mathematics, particularly in mathematical logic and set theory, a club set is a subset of a limit ordinal which is closed under the order topology, and is unbounded relative to the limit ordinal. The name club is a contraction of "closed and unbounded".

Formally, if $\kappa$ is a limit ordinal, then a set $C \subseteq \kappa$ is *closed* in $\kappa$ if and only if for every $\alpha < \kappa$, if $\sup(C \cap \alpha) = \alpha \neq 0$, then $\alpha \in C$. Thus, if the limit of some sequence from $C$ is less than $\kappa$, then the limit is also in $C$.

If $\kappa$ is a limit ordinal and $C \subseteq \kappa$ then $C$ is unbounded in $\kappa$ if for any $\alpha < \kappa$, there is some $\beta \in C$ such that $\alpha < \beta$.

If a set is both closed and unbounded, then it is a club set. Closed proper classes are also of interest (every proper class of ordinals is unbounded in the class of all ordinals).

For example, the set of all countable limit ordinals is a club set with respect to the first uncountable ordinal; but it is not a club set with respect to any higher limit ordinal, since it is neither closed nor unbounded. The set of all limit ordinals $\alpha < \kappa$ is closed unbounded in $\kappa$ ($\kappa$ regular). In fact a club set is nothing else but the range of a normal function (i.e. increasing and continuous).

More generally, if $X$ is a nonempty set and $\lambda$ is a cardinal, then $C \subseteq [X]^{\lambda}$ is *club* if every union of a subset of $C$ is in $C$ and every subset of $X$ of cardinality less than $\lambda$ is contained in some element of C.

## Closed Unbounded Filter

Let $\kappa$ be a limit ordinal of uncountable cofinality $\lambda$. For some $\alpha < \lambda$, let $\langle C_\xi : \xi < \alpha \rangle$ be a sequence of closed unbounded subsets of $\kappa$. Then $\bigcap_{\xi<\alpha} C_\xi$ is also closed unbounded. To see this, one can note that an intersection of closed sets is always closed, so we just need to show that this intersection is unbounded. So fix any $\beta_0 < \kappa$, and for each n<ω choose from each $C_\xi$ an element $\beta^{\xi}_{n+1} > \beta_n$, which is possible because each is unbounded. Since this is a collection of fewer than $\lambda$ ordinals, all less than $\kappa$, their least upper bound must also be less than $\kappa$, so we can call it $\beta_{n+1}$. This process generates a countable sequence $\beta_0, \beta_1, \beta_2, \ldots$. The limit of this sequence must in fact also be the limit of the sequence $\beta^{\xi}_0, \beta^{\xi}_1, \beta^{\xi}_2, \ldots$, and since each $C_\xi$ is closed and $\lambda$ is uncountable, this limit must be in each $C_\xi$, and therefore this limit is an element of the intersection that is above $\beta_0$, which shows that the intersection is unbounded.

From this, it can be seen that if $\kappa$ is a regular cardinal, then $\{S \subset \kappa : \exists C \subset S$ such that $C$ is closed unbounded in $\kappa\}$ is a non-principal $\kappa$-complete filter on $\kappa$.

If $\kappa$ is a regular cardinal then club sets are also closed under diagonal intersection.

In fact, $\kappa$ if is regular and $\mathcal{F}$ is any filter on $\kappa$ closed under diagonal intersection, containing all sets of the form $\{\xi < \kappa : \xi \geq \alpha\}$ for $\alpha < \kappa$, then $\mathcal{F}$ must include all club sets.

# Empty Set

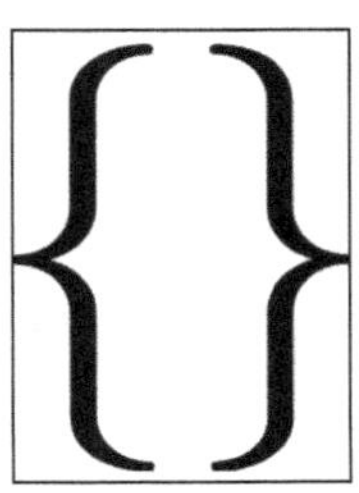

The empty set is the set containing no elements.

In mathematics, the empty set is the unique set having no elements; its size or cardinality (count of elements in a set) is zero. Some axiomatic set theories ensure that the empty set exists by including an axiom of empty set; in other theories, its existence can be deduced. Many possible properties of sets are vacuously true for the empty set.

The empty set is referred to as the "null set". However *null set* is a distinct notion within the context of measure theory. In that setting, it describes a set of measure zero; such a set is not necessarily empty. The empty set may also be called the *void set*. Common notations for the empty set include "{}", "$\emptyset$", and "ø".

## Notation

Common notations for the empty set include "{}", "$\emptyset$", and "ø". The latter two symbols were introduced by the Bourbaki group (specifically André Weil) in 1939, inspired by the letter Ø in the Norwegian and Danish alphabets (and not related in any way to the letter Φ). In the past, "o" was occasionally used as a symbol for the empty set, but this is now considered to be an improper use of notation.

A symbol for the empty set.

The symbol ø is available at Unicode point U+2205. It can be coded in HTML as ∅ and as ∅. It can be coded in LaTeX as \varnothing. The symbol $\emptyset$ is coded in LaTeX as \emptyset.

## Properties

In standard axiomatic set theory, by the principle of extensionality, two sets are equal if they have the same elements; therefore there can be only one set with no elements. Hence there is but one empty set, and we speak of "the empty set" rather than "an empty set".

For any set $A$:

- The empty set is a subset of $A$:

  $\forall A : \emptyset \subseteq A$

- The union of $A$ with the empty set is $A$:

  $\forall A : A \cup \emptyset = A$

- The intersection of $A$ with the empty set is the empty set:

  $\forall A : A \cap \emptyset = \emptyset$

- The Cartesian product of $A$ and the empty set is the empty set:

  $\forall A : A \times \emptyset = \emptyset$

The empty set has the following properties:

- Its only subset is the empty set itself:

  $\forall A : A \subseteq \varnothing \Rightarrow A = \varnothing$

- The power set of the empty set is the set containing only the empty set:

  $2^{\varnothing} = \{\varnothing\}$

- Its number of elements (that is, its cardinality) is zero:

  $|\varnothing| = 0$

The connection between the empty set and zero goes further, however: in the standard set-theoretic definition of natural numbers, sets are used to model the natural numbers. In this context, zero is modelled by the empty set.

For any property:

- For every element of $\varnothing$ the property holds (vacuous truth).
- There is no element of $\varnothing$ for which the property holds.

Conversely, if for some property and some set $V$, the following two statements hold:

- For every element of $V$ the property holds.
- There is no element of $V$ for which the property holds.

Then $V = \emptyset$.

By the definition of subset, the empty set is a subset of any set $A$. That is, *every* element $x$ of $\varnothing$ belongs to $A$. Indeed, if it were not true that every element of $\varnothing$ is in $A$ then there would be at least one element of $\varnothing$ that is not present in $A$. Since there are *no* elements of $\varnothing$ at all, there is no element of $\varnothing$ that is not in $A$. Any statement that begins "for every element of $\varnothing$" is not making any substantive claim; it is a vacuous truth. This is often paraphrased as "everything is true of the elements of the empty set."

## Operations on the Empty Set

When speaking of the sum of the elements of a finite set, one is inevitably led to the convention that the sum of the elements of the empty set is zero. The reason for this is that zero is the identity element for addition. Similarly, the product of the elements of the empty set should be considered to be one, since one is the identity element for multiplication.

A derangement is a permutation of a set without fixed points. The empty set can be considered a derangement of itself, because it has only one permutation ($0! = 1$), and it is vacuously true that no element (of the empty set) can be found that retains its original position.

## In other Areas of Mathematics

### Extended Real Numbers

Since the empty set has no members, when it is considered as a subset of any ordered set, then every member of that set will be an upper bound and lower bound for the empty set. For example, when considered as a subset of the real numbers, with its usual ordering, represented by the real number line, every real number is both an upper and lower bound for the empty set. When considered as a subset of the extended reals formed by adding two "numbers" or "points" to the real numbers, namely negative infinity, denoted $-\infty$ which is defined to be less than every other extended real number, and positive infinity, denoted $+\infty$ which is defined to be greater than every other extended real number, then:

$$\sup\varnothing = \min(\{-\infty, +\infty\} \cup \mathbb{R}) = -\infty,$$

and

$$\inf\varnothing = \max(\{-\infty, +\infty\} \cup \mathbb{R}) = +\infty.$$

That is, the least upper bound (sup or supremum) of the empty set is negative infinity, while the greatest lower bound (inf or infimum) is positive infinity. By analogy with the above, in the domain of the extended reals, negative infinity is the identity element for the maximum and supremum operators, while positive infinity is the identity element for minimum and infimum.

### Topology

In any topological space *X*, the empty set is open by definition, as is *X*. Since the complement of an open set is closed and the empty set and *X* are complements of each other, the empty set is also closed, making it a clopen set. Moreover, the empty set is compact by the fact that every finite set is compact.

The closure of the empty set is empty. This is known as "preservation of nullary unions."

### Category Theory

If *A* is a set, then there exists precisely one function *f* from ∅ to *A*, the empty function. As a result, the empty set is the unique initial object of the category of sets and functions.

The empty set can be turned into a topological space, called the empty space, in just one way: by defining the empty set to be open. This empty topological space is the unique initial object in the category of topological spaces with continuous maps. In fact, it is a strict initial object: only the empty set has a function to the empty set.

### Set Theory

In the von Neumann construction of the ordinals, 0 is defined as the empty set, and the successor of an ordinal is defined as $S(\alpha) = \alpha \cup \{\alpha\}$. Thus, we have $0 = \varnothing$, $1 = 0 \cup \{0\} = \{\varnothing\}$, $2 = 1 \cup \{1\} = \{\varnothing, \{\varnothing\}\}$, and so on. The von Neumann construction, along with the axiom of infinity, which guarantees the existence of at least one infinite set, can be used to construct the set of natural numbers, $\mathbb{N}_0$, such that the Peano axioms of arithmetic are satisfied.

## Power Set

In mathematics, the power set (or powerset) of any set *S* is the set of all subsets of *S*, including the empty set and *S* itself, variously denoted as $P(S)$, $\mathcal{P}(S)$, $\wp(S)$ (using the "Weierstrass p"), $P(S)$, P(*S*), or, identifying the powerset of *S* with the set of all functions from *S* to a given set of two elements, $2^S$. In axiomatic set theory (as developed, for example, in the ZFC axioms), the existence of the power set of any set is postulated by the axiom of power set.

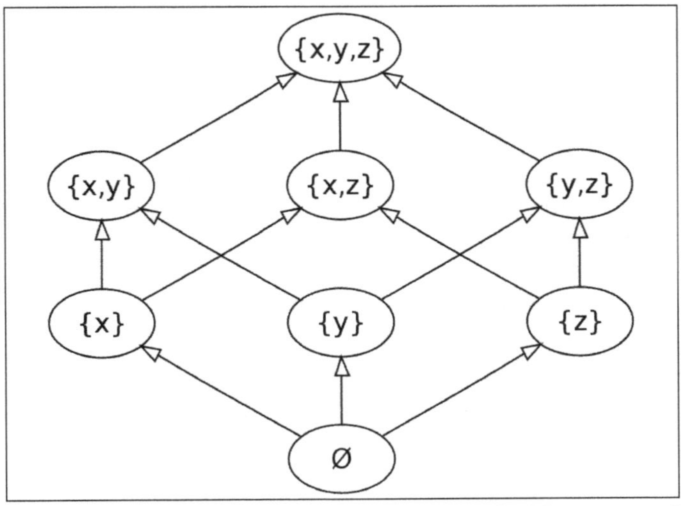


The elements of the power set of the set {*x*, *y*, *z*} ordered with respect to inclusion.

Any subset of P(S) is called a family of sets over S.

Example - If *S* is the set {*x*, *y*, *z*}, then the subsets of *S* are:

- {} (also denoted $\varnothing$ or $\theta$, the empty set or the null set)
- {*x*}
- {*y*}
- {*z*}
- {*x*, *y*}
- {*x*, *z*}
- {*y*, *z*}
- {*x*, *y*, *z*}

and hence the power set of *S* is {{}, {*x*}, {*y*}, {*z*}, {*x*, *y*}, {*x*, *z*}, {*y*, *z*}, {*x*, *y*, *z*}}.

## Properties

If $S$ is a finite set with $|S| = n$ elements, then the number of subsets of $S$ is $|P(S)| = 2^n$. This fact, which is the motivation for the notation $2^S$, may be demonstrated simply as follows:

> "First, order the elements of $S$ in any manner. We write any subset of $S$ in the format $\{\gamma_1, \gamma_2, ..., \gamma_n\}$ where $\gamma_i$, $1 \leq i \leq n$, can take the value of 0 or 1. If $\gamma_i = 1$, the $i$-th element of $S$ is in the subset; otherwise, the $i$-th element is not in the subset. Clearly the number of distinct subsets that can be constructed this way is $2^n$ as $\gamma_i \in \{0, 1\}$".

Cantor's diagonal argument shows that the power set of a set (whether infinite or not) always has strictly higher cardinality than the set itself (informally the power set must be larger than the original set). In particular, Cantor's theorem shows that the power set of a countably infinite set is uncountably infinite. The power set of the set of natural numbers can be put in a one-to-one correspondence with the set of real numbers.

The power set of a set $S$, together with the operations of union, intersection and complement can be viewed as the prototypical example of a Boolean algebra. In fact, one can show that any *finite* Boolean algebra is isomorphic to the Boolean algebra of the power set of a finite set. For *infinite* Boolean algebras this is no longer true, but every infinite Boolean algebra can be represented as a subalgebra of a power set Boolean algebra.

The power set of a set $S$ forms an abelian group when considered with the operation of symmetric difference (with the empty set as the identity element and each set being its own inverse) and a commutative monoid when considered with the operation of intersection. It can hence be shown (by proving the distributive laws) that the power set considered together with both of these operations forms a Boolean ring.

## Representing Subsets as Functions

In set theory, $X^Y$ is the set of all functions from $Y$ to $X$. As "2" can be defined as $\{0,1\}$, $2^S$ (i.e., $\{0,1\}^S$) is the set of all functions from $S$ to $\{0,1\}$. By identifying a function in $2^S$ with the corresponding preimage of 1, we see that there is a bijection between $2^S$ and $P(S)$, where each function is the characteristic function of the subset in $P(S)$ with which it is identified. Hence $2^S$ and $P(S)$ could be considered identical set-theoretically. Thus there are two distinct notational motivations for denoting the power set by $2^S$: the fact that this function-representation of subsets makes it a special case of the $X^Y$ notation and the property, that $|2^{S|} = 2^{|S|}$.

This notion can be applied to the example above in which $S = \{x, y, z\}$ to see the isomorphism with the binary numbers from 0 to $2^n - 1$ with $n$ being the number of elements in the set. In $S$, a "1" in the position corresponding to the location in the enumerated set $\{ (x, 0), (y, 1), (z, 2) \}$ indicates the presence of the element. So $\{x, y\} = 011_{(2)}$.

For the whole power set of $S$ we get:

| Subset | Sequence of digits | Binary interpretation | Decimal equivalent |
|---|---|---|---|
| $\{\}$ | 0, 0, 0 | $000_{(2)}$ | $0_{(10)}$ |
| $\{x\}$ | 0, 0, 1 | $001_{(2)}$ | $1_{(10)}$ |

| $\{y\}$ | 0, 1, 0 | $010_{(2)}$ | $2_{(10)}$ |
|---|---|---|---|
| $\{x, y\}$ | 0, 1, 1 | $011_{(2)}$ | $3_{(10)}$ |
| $\{z\}$ | 1, 0, 0 | $100_{(2)}$ | $4_{(10)}$ |
| $\{x, z\}$ | 1, 0, 1 | $101_{(2)}$ | $5_{(10)}$ |
| $\{y, z\}$ | 1, 1, 0 | $110_{(2)}$ | $6_{(10)}$ |
| $\{x, y, z\}$ | 1, 1, 1 | $111_{(2)}$ | $7_{(10)}$ |

Such bijective mapping of *S* to integers is arbitrary, so this representation of subsets of *S* is not unique, but the sort order of the enumerated set does not change its cardinality.

However, such finite binary representation is only possible if *S* can be enumerated (this is possible even if *S* has an infinite cardinality, such as the set of integers or rationals, but not for example if *S* is the set of real numbers, in which we cannot enumerate all irrational numbers to assign them a defined finite location in an ordered set containing all irrational numbers).

## Relation to Binomial Theorem

The power set is closely related to the binomial theorem. The number of subsets with *k* elements in the power set of a set with *n* elements is given by the number of combinations, C(*n*, *k*), also called binomial coefficients.

For example, the power set of a set with three elements, has:

- C(3, 0) = 1 subset with 0 elements (the empty subset),
- C(3, 1) = 3 subsets with 1 element (the singleton subsets),
- C(3, 2) = 3 subsets with 2 elements (the complements of the singleton subsets),
- C(3, 3) = 1 subset with 3 elements (the original set itself).

Using this relationship we can compute $|2^S|$ using the formula:

$$|2^S| = \sum_{k=0}^{|S|} \binom{|S|}{k}$$

Therefore one can deduce the following identity, assuming $|S| = n$:

$$|2^S| = 2^n = \sum_{k=0}^{n} \binom{n}{k}$$

## Algorithms

If *S* is a finite set, there is a recursive algorithm to calculate *P*(*S*).

Define the operation $F(e, T) = \{X \cup \{e\} \mid X \in T\}$.

In English, return the set with the element *e* added to each set *X* in *T*.

- If $S = \{\}$, then $P(S) = \{\{\}\}$ is returned.
- Otherwise:
  - Let $e$ be any single element of $S$.
  - Let $T = S \setminus \{e\}$ be the set $S$ with element $e$ excluded (a relative complement of the singleton $\{e\}$ in $S$).
  - And the result: $P(S) = P(T) \cup F(e, P(T))$ is returned.

In words: The power set of the empty set is the set containing only the empty set and the power set of any other set is all the subsets of the set containing some specific element and all the subsets of the set not containing that specific element.

The following is a Python implementation of the above algorithm as used for lists (note that such lists, opposed to sets, may also hold duplicated enties):

Verbose:

```
def powerset(s) -> List:
    """Power set."""
    if s == []:  # Base case
        return [s]  # If s is empty, then the only sublist of s is s itself
    else:
        e = s[0]  # Any element from s (in this implementation, we choose the first element)
        t = s[1:]  # s with e removed
        pt = p(t)  # The list of all sublists of t (note that this is a recursive call)
        fept = [x + [e] for x in pt]  # pt with e appended to each sublist
        return pt + fept  # The concatenation of all constructed sublists
# Example:
print(powerset(['x', 'y', 'z']))
```

Concise form:

```
def p(s):
    return p(s[1:]) + [x + [s[0]] for x in p(s[1:])] if s else [s]
```

## Subsets of Limited Cardinality

The set of subsets of $S$ of cardinality less than or equal to $\kappa$ is sometimes denoted by $P_\kappa(S)$ or $[S]^\kappa$, and the set of subsets with cardinality strictly less than $\kappa$ is sometimes denoted $P_{<\kappa}(S)$ or $[S]^{<\kappa}$. Similarly, the set of non-empty subsets of $S$ might be denoted by $P_{\geq 1}(S)$ or $P^+(S)$.

### Power Object

A set can be regarded as an algebra having no nontrivial operations or defining equations. From this perspective the idea of the power set of $X$ as the set of subsets of $X$ generalizes naturally to the subalgebras of an algebraic structure or algebra.

Now the power set of a set, when ordered by inclusion, is always a complete atomic Boolean algebra, and every complete atomic Boolean algebra arises as the lattice of all subsets of some set. The generalization to arbitrary algebras is that the set of subalgebras of an algebra, again ordered by inclusion, is always an algebraic lattice, and every algebraic lattice arises as the lattice of subalgebras of some algebra. So in that regard subalgebras behave analogously to subsets.

However, there are two important properties of subsets that do not carry over to subalgebras in general. First, although the subsets of a set form a set (as well as a lattice), in some classes it may not be possible to organize the subalgebras of an algebra as itself an algebra in that class, although they can always be organized as a lattice. Secondly, whereas the subsets of a set are in bijection with the functions from that set to the set $\{0,1\} = 2$, there is no guarantee that a class of algebras contains an algebra that can play the role of 2 in this way.

Certain classes of algebras enjoy both of these properties. The first property is more common, the case of having both is relatively rare. One class that does have both is that of multigraphs. Given two multigraphs $G$ and $H$, a homomorphism $h$: $G \to H$ consists of two functions, one mapping vertices to vertices and the other mapping edges to edges. The set $H^G$ of homomorphisms from $G$ to $H$ can then be organized as the graph whose vertices and edges are respectively the vertex and edge functions appearing in that set. Furthermore, the subgraphs of a multigraph $G$ are in bijection with the graph homomorphisms from $G$ to the multigraph $\Omega$ definable as the complete directed graph on two vertices (hence four edges, namely two self-loops and two more edges forming a cycle) augmented with a fifth edge, namely a second self-loop at one of the vertices. We can therefore organize the subgraphs of $G$ as the multigraph $\Omega^G$, called the power object of $G$.

What is special about a multigraph as an algebra is that its operations are unary. A multigraph has two sorts of elements forming a set $V$ of vertices and $E$ of edges, and has two unary operations $s,t$: $E \to V$ giving the source (start) and target (end) vertices of each edge. An algebra all of whose operations are unary is called a presheaf. Every class of presheaves contains a presheaf $\Omega$ that plays the role for subalgebras that 2 plays for subsets. Such a class is a special case of the more general notion of elementary topos as a category that is closed (and moreover cartesian closed) and has an object $\Omega$, called a subobject classifier. Although the term "power object" is sometimes used synonymously with exponential object $Y^X$, in topos theory $Y$ is required to be $\Omega$.

## Binary Relation

In mathematics, a binary relation over two sets $A$ and $B$ is a set of ordered pairs $(a, b)$ consisting of elements $a$ of $A$ and elements $b$ of $B$. That is, it is a subset of the Cartesian product $A \times B$. It encodes the information of relation: an element $a$ is related to an element $b$ if and only if the pair $(a, b)$ belongs to the set.

An example is the "divides" relation over the set of prime numbers P and the set of integers Z, in which each prime $p$ is related to each integer $z$ that is a multiple of $p$, but not to an integer that is not a multiple of $p$. In this relation, for instance, the prime 2 is related to numbers such as −4, 0, 6, 10, but not 1 or 9, and the prime 3 is related to 0, 6, and 9, but not 4 or 13.

Binary relations are used in many branches of mathematics to model concepts like "is greater than", "is equal to", and "divides" in arithmetic, "is congruent to" in geometry, "is adjacent to" in graph theory, "is orthogonal to" in linear algebra and many more. A function may be defined as a special kind of binary relation. Binary relations are also heavily used in computer science.

A binary relation is the special case $n = 2$ of an $n$-ary relation $R \subseteq A_1 \times \cdots \times A_n$, that is, a set of $n$-tuples where the $j$th component of each $n$-tuple is taken from the $j$th domain $A_j$ of the relation. An example of a ternary relation over Z is "... lies between ... and ...", which contains triples such as (5, 2, 8), (5, 8, 2), and (−4, 9, −7).

A binary relation over $A$ and $B$ is an element of the power set of $A \times B$. Since the latter set is ordered by inclusion (⊆), each relation has a place in the lattice of subsets of $A \times B$.

As part of set theory, relations are manipulated with the algebra of sets, including complementation. Furthermore, the two sets are considered symmetrically by introduction of the converse relation which exchanges their places. Another operation is composition of relations. Altogether these tools form the calculus of relations A deeper analysis of relations involves decomposing them into subsets called concepts and placing them in a complete lattice.

In some systems of axiomatic set theory, relations are extended to classes, which are generalizations of sets. This extension is needed for, among other things, modeling the concepts of "is an element of" or "is a subset of" in set theory, without running into logical inconsistencies such as Russell's paradox.

The terms correspondence, dyadic relation and two-place relation are synonyms for binary relation. But some authors use the term "binary relation" for any subset of a Cartesian product $A \times B$ without reference to $A$ and $B$ while the term "correspondence" is reserved for a binary relation with reference to $A$ and $B$.

Given two sets $X$ and $Y$, the Cartesian product $X \times Y$ is defined as $\{(x, y) \mid x \in X \text{ and } y \in Y\}$, and its elements are called ordered pairs.

A binary relation $R$ on $X$ and $Y$ is a subset of $X \times Y$; that is, it is a set of ordered pairs $(x, y)$ consisting of elements $x \in X$ and $y \in Y$. The set $X$ is called the *set of departure* and the set $Y$ the *set of destination* or *codomain*. (In order to specify the choices of the sets $X$ and $Y$, some authors define a binary relation or a correspondence as an ordered triple $(X, Y, R)$ where $R$ is a subset of $X \times Y$.) The statement $(x, y) \in R$ is read "$x$ is $R$-related to $y$", and is denoted by $xRy$.

When $X = Y$, a binary relation is called a homogeneous relation. To emphasize the fact $X$ and $Y$ are allowed to be different, a binary relation is also called a heterogeneous relation. An example of a homogeneous relation is a kinship where the relations are over people. Homogeneous relation may be viewed as directed graphs, and in the symmetric case as ordinary graphs. Homogeneous relations also encompass orderings as well as partitions of a set (called equivalence relations).

The order of the elements is important; if $a \neq b$ then $aRb$ and $bRa$ can be true or false independently of each other. For example, 3 divides 9, but 9 does not divide 3.

The domain of $R$ is the set of all $x$ such that $xRy$ for at least one $y$. The range or *image* of $R$ is the set of all $y$ such that $xRy$ for at least one $x$. The *field* of $R$ is the union of its domain and its range.

Some authors also call a binary relation a multivalued function; in fact, a (single-valued) partial function from $X$ to $Y$ is nothing but a binary relation over $X$ and $Y$ such that $xRy$ and $xRz \Rightarrow y = z$ for all $x$ in $X$ and $y$, $z$ in $Y$.

## Example

| 1st example relation | | | | |
|---|---|---|---|---|
| | ball | car | doll | cup |
| John | + | – | – | – |
| Mary | – | – | + | – |
| Ian | – | – | – | – |
| Venus | – | + | – | – |

| 2nd example relation | | | | |
|---|---|---|---|---|
| | ball | car | doll | cup |
| John | + | – | – | – |
| Mary | – | – | + | – |
| Venus | – | + | – | – |

The following example shows that the choice of codomain is important. Suppose there are four objects $A$ = {ball, car, doll, cup} and four people $B$ = {John, Mary, Ian, Venus}. A possible relation on $A$ and $B$ is "is owned by", given by $R$ = {(ball, John), (doll, Mary), (car, Venus)}. That is, John owns the ball, Mary owns the doll, and Venus owns the car. Nobody owns the cup and Ian owns nothing. As a set, $R$ does not involve Ian, and therefore $R$ could have been viewed as a subset of $A$ × {John, Mary, Venus}, i.e. a relation over $A$ and {John, Mary, Venus}.

## Special Types of Binary Relations

Example relations over the real numbers.

$y = x^2$.

$y = 2x + 20$.

Some important types of binary relations $R$ over two sets $X$ and $Y$ are listed below.

Uniqueness properties:

- Injective (also called left-unique): For all $x$ and $z$ in $X$ and $y$ in $Y$, if $xRy$ and $zRy$ then $x = z$. For example, the green relation in the diagram is injective, but the red relation is not, as, e.g., it relates both −5 and 5 to 25.

- Functional (also called right-unique, right-definite or univalent): For all $x$ in $X$, and $y$ and $z$ in $Y$, if $xRy$ and $xRz$ then $y = z$; such a binary relation is called a *partial function*. Both relations in the picture are functional. An example of a non-functional relation can be obtained by rotating the red graph clockwise by 90 degrees, i.e. by considering the relation $x = y^2$ which, e.g., relates 25 to both −5 and 5.

- One-to-one (also written 1-to-1): Injective and functional. The green relation is one-to-one, but the red is not.

Totality properties (only definable if the sets of departure $X$ resp. destination $Y$ are specified):

- Left-total: For all $x$ in $X$ there exists a $y$ in $Y$ such that $xRy$; such an $R$ is also called a multivalued function by some authors. This property, although also referred to as *total* by some authors, is different from the definition of *total*. Both relations in the picture are left-total. The relation $x = y^2$, obtained from the above rotation, is not left-total, as, e.g., it doesn't relate −14 to any real number.

- Surjective (also called right-total or onto): For all $y$ in $Y$ there exists an $x$ in $X$ such that $xRy$. The green relation is surjective, but the red relation is not, as, e.g., it doesn't relate any real number to −14.

Uniqueness and totality properties:

- A function: A relation that is functional and left-total. Both the green and the red relation are functions.

- An injective function or injection: A relation that is injective, functional, and left-total.

- A surjective function or surjection: A relation that is functional, left-total, and right-total.

- A bijection: A surjective one-to-one or surjective injective function is said to be bijective, also known as one-to-one correspondence. The green relation is bijective, but the red is not.

## Operations on Binary Relations

If $R$ and $S$ are binary relations over two sets $X$ and $Y$ then each of the following is a binary relation over $X$ and $Y$:

- Union: $R \cup S \subseteq X \times Y$, defined as $R \cup S = \{(x, y) \mid (x, y) \in R \text{ or } (x, y) \in S\}$. The identity element is the empty relation. For example, $\geq$ is the union of $>$ and $=$.

- Intersection: $R \cap S \subseteq X \times Y$, defined as $R \cap S = \{(x, y) \mid (x, y) \in R \text{ and } (x, y) \in S\}$. The identity element is the universal relation.

If $R$ is a binary relation over $X$ and $Y$, and $S$ is a binary relation over $Y$ and $Z$ then the following is a binary relation over $X$ and $Z$:

- Composition: $S \circ R$, also denoted $R ; S$ (or $R \circ S$), defined as $S \circ R = \{(x, z) \mid \exists y \in Y, (x, y) \in R \text{ and } (y, z) \in S\}$. The identity element is the identity relation. The order of $R$ and $S$ in the notation $S \circ R$, used here agrees with the standard notational order for composition of functions. For example, the composition "is mother of" $\circ$ "is parent of" yields "is maternal grandparent of", while the composition "is parent of" $\circ$ "is mother of" yields "is grandmother of".

A relation $R$ over two sets $X$ and $Y$ is said to be *contained* in a relation $S$ over $X$ and $Y$ if $R$ is a subset of $S$, that is, for all $x$ in $X$ and $y$ in $Y$, if $xRy$ then $xSy$. In this case, if $R$ and $S$ disagree, $R$ is also said to be *smaller* than $S$. For example, $>$ is contained in $\geq$.

If $R$ is a binary relation over $X$ and $Y$ then the following is a binary relation over $Y$ and $X$:

- Converse: $R^{\mathrm{T}}$, defined as $R^{\mathrm{T}} = \{(y, x) \mid (x, y) \in R\}$. A binary relation over a set is equal to its converse if and only if it is symmetric. For example, "is less than" ($<$) is the converse of "is greater than" ($>$).

## Complement

If $R$ is a binary relation in $X \times Y$ then it has a *complementary relation* $S$ defined as $xSy \Leftrightarrow \neg xRy$.

An overline or bar is used to indicate the complementary relation: $S = \overline{R}$. Alternatively, a strike-through is used to denote complements, for example, $=$ and $\neq$ are complementary to each other, as are $\in$ and $\notin$, and $\supseteq$ and $\not\supseteq$. Some authors even use $R$ and $\not R$. In total orderings $<$ and $\geq$ are complements, as are $>$ and $\leq$.

The complement of the converse relation $R^{\mathrm{T}}$ is the converse of the complement: $\overline{R^{\mathrm{T}}} = \overline{R}^{\mathrm{T}}$.

If $X = Y$, the complement has the following properties:

- If a relation is symmetric, the complement is too.
- The complement of a reflexive relation is irreflexive and vice versa.
- The complement of a strict weak order is a total preorder and vice versa.

## Restriction

The restriction of a binary relation over a set $X$ to a subset $S$ is the set of all pairs $(x, y)$ in the relation for which $x$ and $y$ are in $S$.

If a relation is reflexive, irreflexive, symmetric, antisymmetric, asymmetric, transitive, total, trichotomous, a partial order, total order, strict weak order, total preorder (weak order), or an equivalence relation, its restrictions are too.

However, the transitive closure of a restriction is a subset of the restriction of the transitive closure, i.e., in general not equal. For example, restricting the relation "$x$ is parent of $y$" to females yields the relation "$x$ is mother of the woman $y$"; its transitive closure doesn't relate a woman with her paternal grandmother. On the other hand, the transitive closure of "is parent of" is "is ancestor of"; its restriction to females does relate a woman with her paternal grandmother.

Also, the various concepts of completeness (not to be confused with being "total") do not carry over to restrictions. For example, over the real numbers a property of the relation $\leq$ is that every non-empty subset $S$ of R with an upper bound in R has a least upper bound (also called supremum) in R. However, for the rational numbers this supremum is not necessarily rational, so the same property does not hold on the restriction of the relation $\leq$ to the rational numbers.

The *left-restriction* (*right-restriction*, respectively) of a binary relation over two sets $X$ and $Y$ to a subset $S$ of its domain (codomain) is the set of all pairs $(x, y)$ in the relation for which $x$ ($y$) is an element of $S$.

## Matrix Representation

Binary relations over two sets $X$ and $Y$ can be represented algebraically by matrices indexed by $X$ and $Y$ with entries in the Boolean semiring (addition corresponds to OR and multiplication to AND), matrix addition corresponds to union of relations, matrix multiplication corresponds to composition of relations (of a relation over $X$ and $Y$ and a relation over $Y$ and $Z$), the Hadamard product corresponds to intersection of relations, the zero matrix corresponds to the empty relation, and the matrix of ones corresponds to the universal relation. If $X$ equals $Y$ then the endorelations form a matrix semiring (indeed, a matrix semialgebra over the Boolean semiring), and the identity matrix corresponds to the identity relation.

## Sets versus Classes

Certain mathematical "relations", such as "equal to", "subset of", and "member of", cannot be understood to be binary relations as defined above, because their domains and codomains cannot be taken to be sets in the usual systems of axiomatic set theory. For example, if we try to model the general concept of "equality" as a binary relation =, we must take the domain and codomain to be the "class of all sets", which is not a set in the usual set theory.

In most mathematical contexts, references to the relations of equality, membership and subset are harmless because they can be understood implicitly to be restricted to some set in the context. The usual work-around to this problem is to select a "large enough" set $A$, that contains all the objects of interest, and work with the restriction $=_A$ instead of =. Similarly, the "subset of" relation $\subseteq$ needs to be restricted to have domain and codomain P($A$) (the power set of a specific set $A$): the resulting set relation can be denoted $\subseteq_A$. Also, the "member of" relation needs to be restricted to have domain $A$ and codomain P($A$) to obtain a binary relation $\in_A$ that is a set. Bertrand Russell has shown that assuming $\in$ to be defined over all sets leads to a contradiction in naive set theory.

Another solution to this problem is to use a set theory with proper classes, such as NBG or Morse–Kelley set theory, and allow the domain and codomain (and so the graph) to be proper classes: in such a theory, equality, membership, and subset are binary relations without special comment. (A minor modification needs to be made to the concept of the ordered triple $(X, Y, G)$, as normally a proper class cannot be a member of an ordered tuple; or of course one can identify the binary relation with its graph in this context.) With this definition one can for instance define a binary relation over every set and its power set.

## Homogeneous Relation

A homogeneous relation (also called endorelation) over a set $X$ is a binary relation over the set $X$ and itself, i.e. it is a subset of the Cartesian product $X \times X$. It is also simply called a binary relation over $X$.

A homogeneous relation $R$ over a set $X$ may be identified with a directed simple graph permitting loops, where $X$ is the vertex set and $R$ is the edge set (there is an edge from a vertex $x$ to a vertex $y$ if and only if $xRy$). The homogenous relation is called the adjacency relation of the directed graph.

The set of all binary relations $\mathcal{B}(X)$ over a set $X$ is the set $2^{X \times X}$ which is a Boolean algebra augmented with the involution of mapping of a relation to its converse relation. Considering composition of relations as a binary operation on $\mathcal{B}(X)$ it forms a inverse semigroup.

## Particular Homogeneous Relations

Some important particular binary relations over a set $X$ are:

- The empty relation $E = \emptyset \subseteq X \times X$,
- The universal relation $U = X \times X$, and
- The identity relation $I = \{(x, x) \mid x \in X\}$.

For arbitrary elements $x$ and $y$ of $X$,

- $xEy$ holds never,
- $xUy$ holds always, and
- $xIy$ holds if and only if $x = y$.

## Properties

Some important properties that a binary relation $R$ over a set $X$ may have are:

- Reflexive: For all $x$ in $X$, $xRx$. For example, $\geq$ is a reflexive relation but $>$ is not.
- Irreflexive (or strict): For no $x$ in $X$, $xRx$. For example, $>$ is an irreflexive relation, but $\geq$ is not.
- Coreflexive: For all $x$ and $y$ in $X$, if $xRy$ then $x = y$. For example, the relation over the integers in which each odd number is related to itself is a coreflexive relation. The equality relation is the only example of a both reflexive and coreflexive relation, and any coreflexive relation is a subset of the identity relation.
- Quasi-reflexive: For all $x$ and $y$ in $X$, if $xRy$ then $xRx$ and $yRy$.

The previous 4 alternatives are far from being exhaustive; e.g., the red relation $y = x^2$ from the above picture is neither irreflexive, nor coreflexive, nor reflexive, since it contains the pair (0, 0), and (2, 4), but not (2, 2), respectively. The latter two facts also rule out quasi-reflexivity.

- Symmetric: For all $x$ and $y$ in $X$, if $xRy$ then $yRx$. For example, "is a blood relative of" is a symmetric relation, because $x$ is a blood relative of $y$ if and only if $y$ is a blood relative of $x$.
- Antisymmetric: For all $x$ and $y$ in $X$, if $xRy$ and $yRx$ then $x = y$. For example, $\geq$ is an antisymmetric relation; so is $>$, but vacuously (the condition in the definition is always false).

- Asymmetric: For all $x$ and $y$ in $X$, if $xRy$ then not $yRx$. A relation is asymmetric if and only if it is both antisymmetric and irreflexive. For example, > is an asymmetric relation, but ≥ is not.

Again, the previous 3 alternatives are far from being exhaustive; as an example over the natural numbers, the relation $xRy$ defined by $x > 2$ is neither symmetric nor antisymmetric, let alone asymmetric.

- Transitive: For all $x$, $y$ and $z$ in $X$, if $xRy$ and $yRz$ then $xRz$. A transitive relation is irreflexive if and only if it is asymmetric. For example, "is ancestor of" is a transitive relation, while "is parent of" is not.
- Connex: For all $x$ and $y$ in $X$, $xRy$ or $yRx$ (or both). This property is sometimes called "total", which is distinct from the definitions of "total".
- Trichotomous: For all $x$ and $y$ in $X$, exactly one of $xRy$, $yRx$ or $x = y$ holds. For example, > is a trichotomous relation, while the relation "divides" over the natural numbers is not.
- Right Euclidean (or just Euclidean): For all $x$, $y$ and $z$ in $X$, if $xRy$ and $xRz$ then $yRz$. For example, = is an Euclidean relation because if $x = y$ and $x = z$ then $y = z$.
- Left Euclidean: For all $x$, $y$ and $z$ in $X$, if $yRx$ and $zRx$ then $yRz$.
- Serial: For all $x$ in $X$, there exists $y$ in $X$ such that $xRy$. For example, > is a serial relation over the integers. But it is not a serial relation over the positive integers, because there is no $y$ in the positive integers such that $1 > y$. However, < is a serial relation over the positive integers, the rational numbers and the real numbers. Every reflexive relation is serial: for a given $x$, choose $y = x$.
- Set-like (or local): For all $x$ in $X$, the class of all $y$ such that $yRx$ is a set. (This makes sense only if relations over proper classes are allowed.) For example, the usual ordering < over the class of ordinal numbers is a set-like relation, while its inverse > is not.
- Well-founded: Every nonempty subset $S$ of $X$ contains a minimal element with respect to $R$. Well-foundedness implies the descending chain condition (that is, no infinite chain ... $x_nR...Rx_3Rx_2Rx_1$ can exist). If the axiom of dependent choice is assumed, both conditions are equivalent.

A preorder is a relation that is reflexive and transitive. A *total preorder*, also called *weak order*, is a relation that is reflexive, transitive, and connex. A *partial order* is a relation that is reflexive, antisymmetric, and transitive. A *total order*, also called *linear order*, *simple order*, *connex order*, or *chain* is a relation that is reflexive, antisymmetric, transitive and connex.

A partial equivalence relation is a relation that is symmetric and transitive. An *equivalence relation* is a relation that is reflexive, symmetric, and transitive. It is also a relation that is symmetric, transitive, and serial, since these properties imply reflexivity.

## Operations on Homogeneous Relations

If $R$ is a homogeneous relation over $X$ then each of the following is a homogeneous relation over $X$:

- Reflexive closure: $R^{=}$, defined as $R^{=} = \{(x, x) \mid x \in X\} \cup R$ or the smallest reflexive relation over $X$ containing $R$. This can be proven to be equal to the intersection of all reflexive relations containing $R$.
- Reflexive reduction: $R^{\neq}$, defined as $R^{\neq} = R \backslash \{(x, x) \mid x \in X\}$ or the largest irreflexive relation over $X$ contained in $R$.
- Transitive closure: $R^{+}$, defined as the smallest transitive relation over $X$ containing $R$. This can be seen to be equal to the intersection of all transitive relations containing $R$.
- Reflexive transitive closure: $R^{*}$, defined as $R^{*} = (R^{+})^{=}$, the smallest preorder containing $R$.
- Reflexive transitive symmetric closure: $R^{\equiv}$, defined as the smallest equivalence relation over $X$ containing $R$.

| Binary endorelations by property | | | | | |
|---|---|---|---|---|---|
| | Reflexivity | Symmetry | Transitivity | Symbol | Example |
| Directed graph | | | | → | |
| Undirected graph | | Symmetric | | | |
| Tournament | Irreflexive | Antisymmetric | | | Pecking order |
| Dependency | Reflexive | Symmetric | | | |
| Preorder | Reflexive | | Yes | ≤ | Preference |
| Strict preorder | Irreflexive | | Yes | < | |
| Total preorder | Reflexive | | Yes | ≤ | |
| Partial order | Reflexive | Antisymmetric | Yes | ≤ | Subset |
| Strict partial order | Irreflexive | Antisymmetric | Yes | < | Proper subset |
| Strict weak order | Irreflexive | Antisymmetric | Yes | < | |
| Total order | Reflexive | Antisymmetric | Yes | ≤ | |
| Partial equivalence relation | | Symmetric | Yes | | |
| Equivalence relation | Reflexive | Symmetric | Yes | ∼, ≅, ≈, ≡ | Equality |

## Number of Homogeneous Relations

The number of distinct binary relations over an $n$-element set is $2^{n^2}$ (sequence A002416 in the OEIS):

| Number of n-element binary relations of different types | | | | | | | | |
|---|---|---|---|---|---|---|---|---|
| Elements | Any | Transitive | Reflexive | Preorder | Partial order | Total preorder | Total order | Equivalence relation |
| 0 | 1 | 1 | 1 | 1 | 1 | 1 | 1 | 1 |
| 1 | 2 | 2 | 1 | 1 | 1 | 1 | 1 | 1 |
| 2 | 16 | 13 | 4 | 4 | 3 | 3 | 2 | 2 |
| 3 | 512 | 171 | 64 | 29 | 19 | 13 | 6 | 5 |
| 4 | 65,536 | 3,994 | 4,096 | 355 | 219 | 75 | 24 | 15 |

| n | $2n^2$ | | $2^{n^2-n}$ | | | $\sum_{k=0}^{n} k!S(n,k)$ | n! | $\sum_{k=0}^{n} S(n,k)$ |
|---|---|---|---|---|---|---|---|---|
| OEIS | A002416 | A006905 | A053763 | A000798 | A001035 | A000670 | A000142 | A000110 |

Notes:

- The number of irreflexive relations is the same as that of reflexive relations.
- The number of strict partial orders (irreflexive transitive relations) is the same as that of partial orders.
- The number of strict weak orders is the same as that of total preorders.
- The total orders are the partial orders that are also total preorders. The number of preorders that are neither a partial order nor a total preorder is, therefore, the number of preorders, minus the number of partial orders, minus the number of total preorders, plus the number of total orders: 0, 0, 0, 3, and 85, respectively.
- The number of equivalence relations is the number of partitions, which is the Bell number.

The binary relations can be grouped into pairs (relation, complement), except that for $n = 0$ the relation is its own complement. The non-symmetric ones can be grouped into quadruples (relation, complement, inverse, inverse complement).

## Examples of Common Homogeneous Relations

- Order relations, including strict orders:
  - Greater than,
  - Greater than or equal to,
  - Less than,
  - Less than or equal to,
  - Divides (evenly),
  - Subset of.
- Equivalence relations:
  - Equality,
  - Parallel with (for affine spaces),
  - Is in bijection with,
  - Isomorphic.
- Tolerance relation, a reflexive and symmetric relation:
  - Dependency relation, a finite tolerance relation.
  - Independency relation, the complement of some dependency relation.
- Kinship relations.

# Set Operations

## Union

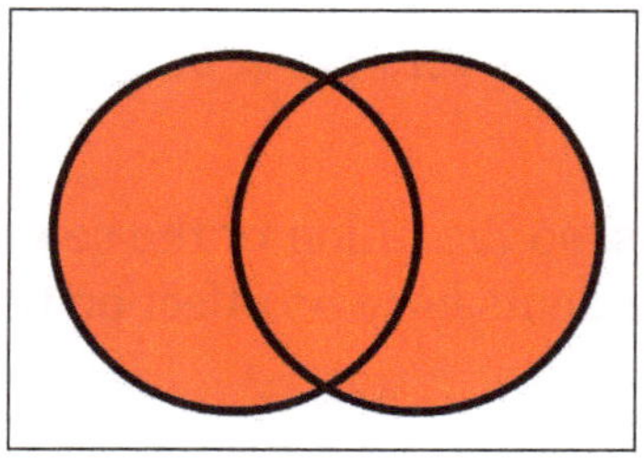

Union of two sets: A ∪ B.

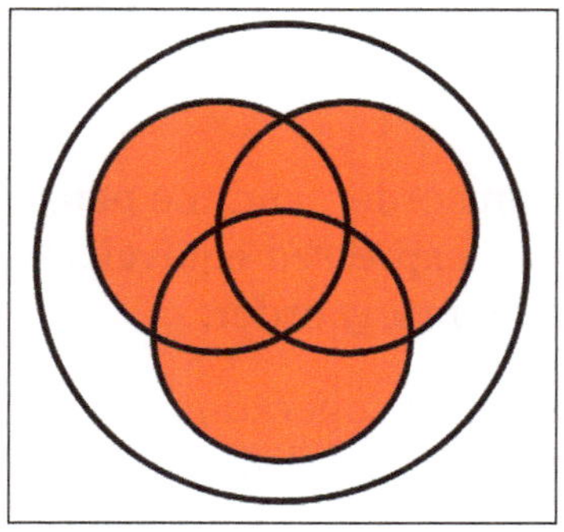

Union of three sets: A ∪ B ∪ C.

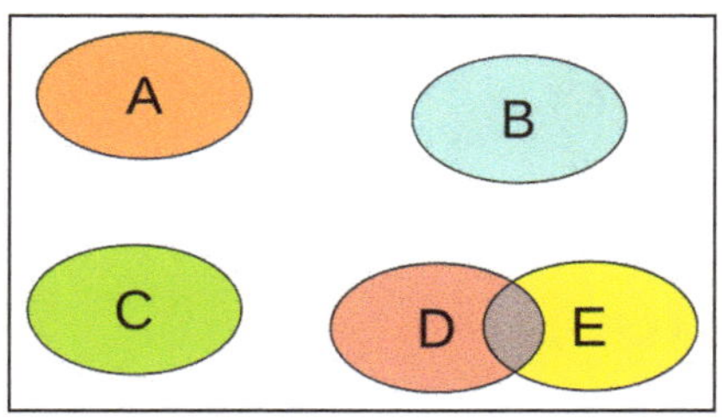


The union of A, B, C, D, and E is everything except the white area.

In set theory, the union (denoted by ∪) of a collection of sets is the set of all elements in the collection. It is one of the fundamental operations through which sets can be combined and related to each other.

### Union of two Sets

The union of two sets $A$ and $B$ is the set of elements which are in $A$, in $B$, or in both $A$ and $B$. In symbols,

$$A \cup B = \{x : x \in A \text{ or } x \in B\}.$$

For example, if $A = \{1, 3, 5, 7\}$ and $B = \{1, 2, 4, 6\}$ then $A \cup B = \{1, 2, 3, 4, 5, 6, 7\}$. A more elaborate example (involving two infinite sets) is:

$$A = \{x \text{ is an even integer larger than } 1\}$$

$B = \{x \text{ is an odd integer larger than } 1\}$

$$A \cup B = \{2,3,4,5,6,\dots\}$$

As another example, the number 9 is *not* contained in the union of the set of prime numbers {2, 3, 5, 7, 11, ...} and the set of even numbers {2, 4, 6, 8, 10, ...}, because 9 is neither prime nor even.

Sets cannot have duplicate elements, so the union of the sets {1, 2, 3} and {2, 3, 4} is {1, 2, 3, 4}. Multiple occurrences of identical elements have no effect on the cardinality of a set or its contents.

## Algebraic Properties

Binary union is an associative operation; that is, for any sets *A*, *B*, and *C*,

$$A \cup (B \cup C) = (A \cup B) \cup C.$$

The operations can be performed in any order, and the parentheses may be omitted without ambiguity (i.e., either of the above can be expressed equivalently as $A \cup B \cup C$). Similarly, union is commutative, so the sets can be written in any order.

The empty set is an identity element for the operation of union. That is, $A \cup \emptyset = A$, for any set *A*. This follows from analogous facts about logical disjunction.

Since sets with unions and intersections form a Boolean algebra, intersection distributes over union,

$$A \cap (B \cup C) = (A \cap B) \cup (A \cap C)$$

and union distributes over intersection,

$$A \cup (B \cap C) = (A \cup B) \cap (A \cup C).$$

Within a given universal set, union can be written in terms of the operations of intersection and complement as,

$$A \cup B = \left(A^C \cap B^C\right)^C$$

where the superscript $^C$ denotes the complement with respect to the universal set.

## Finite Unions

One can take the union of several sets simultaneously. For example, the union of three sets *A*, *B*, and *C* contains all elements of *A*, all elements of *B*, and all elements of *C*, and nothing else. Thus, *x* is an element of $A \cup B \cup C$ if and only if *x* is in at least one of *A*, *B*, and *C*.

A finite union is the union of a finite number of sets; the phrase does not imply that the union set is a finite set.

## Arbitrary Unions

The most general notion is the union of an arbitrary collection of sets, sometimes called an *infinitary union*. If M is a set or class whose elements are sets, then $x$ is an element of the union of M if and only if there is at least one element $A$ of M such that $x$ is an element of $A$. In symbols:

$$x \in \bigcup \mathbf{M} \Leftrightarrow \exists A \in \mathbf{M}, x \in A.$$

This idea subsumes the preceding sections—for example, $A \cup B \cup C$ is the union of the collection $\{A, B, C\}$. Also, if M is the empty collection, then the union of M is the empty set.

## Notations

The notation for the general concept can vary considerably. For a finite union of sets $S_1, S_2, S_3, \ldots, S_n$ one often writes $S_1 \cup S_2 \cup S_3 \cup \ldots \cup S_n$ or $\bigcup_{i=1}^{n} S_i$. Various common notations for arbitrary unions include $\bigcup \mathbf{M}$, $\bigcup_{A \in \mathbf{M}} A$, and $\bigcup_{i \in I} A_i$, the last of which refers to the union of the collection $\{A_i : i \in I\}$ where $I$ is an index set and $A_i$ is a set for every $i \in I$. In the case that the index set $I$ is the set of natural numbers, one uses a notation $\bigcup_{i=1}^{\infty} A_i$ analogous to that of the infinite sums in series.

When the symbol "∪" is placed before other symbols instead of between them, it is usually rendered as a larger size.

## Complement

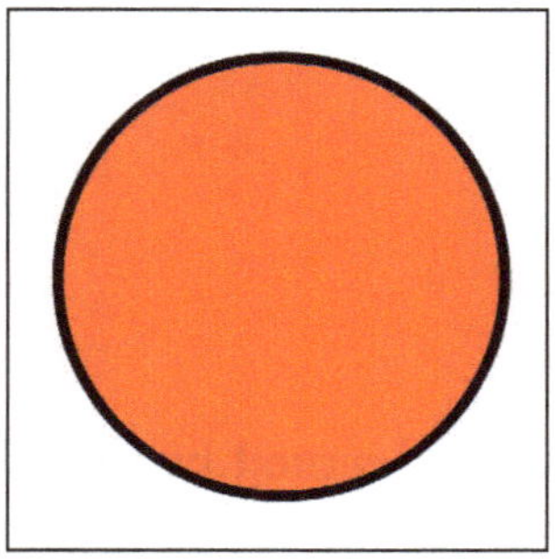

If $A$ is the area colored red.

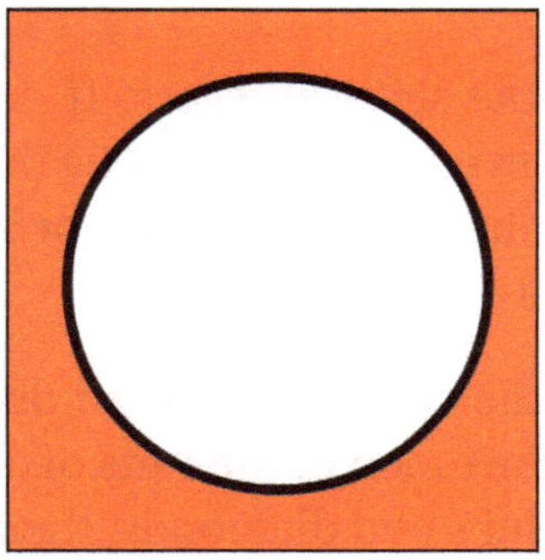

Then the complement of $A$ is everything else.

In set theory, the complement of a set *A* refers to elements not in *A*.

When all sets under consideration are considered to be subsets of a given set *U*, the absolute complement of *A* is the set of elements in *U* but not in *A*.

The relative complement of *A* with respect to a set *B*, also termed the difference of sets *A* and *B*, written $B \setminus A$, is the set of elements in *B* but not in *A*.

## Absolute Complement

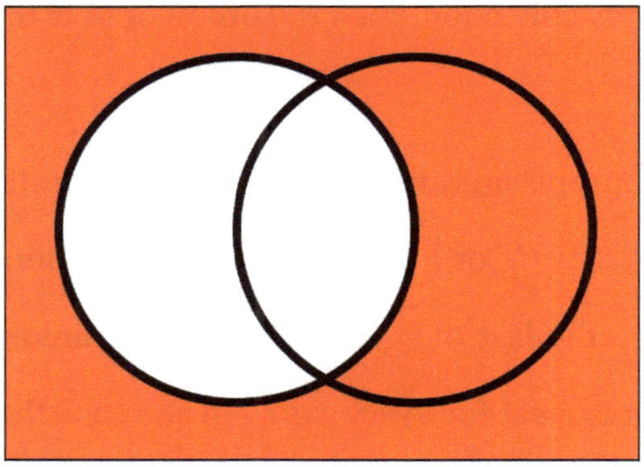

The absolute complement of *A* (left circle) in *U*: $A^c = U \setminus A$.

If *A* is a set, then the absolute complement of *A* (or simply the complement of A) is the set of elements not in *A*, within a larger set that is implicitely defined. In other words, let *U* be a set that contains all the elements under study; if it is no need to mention *U*, either because it has been previously specified, or it is obvious and unique, then the absolute complement of *A* is the relative complement of *A* in *U*:

$$A^c = U \setminus A.$$

Formally:

$$A^c = \{x \in U \mid x \notin A\}$$

The absolute complement of *A* is usually denoted by $A^c$. Other notations include $\overline{A}$, $A'$, $\complement_U A$, and $\complement A$.

Examples:

- Assume that the universe is the set of integers. If *A* is the set of odd numbers, then the complement of *A* is the set of even numbers. If *B* is the set of multiples of 3, then the complement of *B* is the set of numbers congruent to 1 or 2 modulo 3 (or, in simpler terms, the integers that are not multiples of 3).
- Assume that the universe is the standard 52-card deck. If the set *A* is the suit of spades, then the complement of *A* is the union of the suits of clubs, diamonds, and hearts. If the set *B* is the union of the suits of clubs and diamonds, then the complement of *B* is the union of the suits of hearts and spades.

## Properties

Let $A$ and $B$ be two sets in a universe $U$. The following identities capture important properties of absolute complements:

De Morgan's laws:

- $(A \cup B)^c = A^c \cap B^c$.
- $(A \cap B)^c = A^c \cup B^c$.

Complement laws:

- $A \cup A^c = U$.
- $A \cap A^c = \varnothing$.
- $\varnothing^c = U$.
- $U^c = \varnothing$.
- If $A \subseteq B$, then $B^c \subseteq A^c$.

(this follows from the equivalence of a conditional with its contrapositive).

Involution or double complement law:

- $(A^c)^c = A$.

Relationships between relative and absolute complements:

- $A \setminus B = A \cap B^c$.
- $(A \setminus B)^c = A^c \cup B$.

Relationship with set difference:

- $A^c \setminus B^c = B \setminus A$.

The first two complement laws above show that if $A$ is a non-empty, proper subset of $U$, then $\{A, A^c\}$ is a partition of $U$.

## Relative Complement

If $A$ and $B$ are sets, then the relative complement of $A$ in $B$, also termed the set difference of $B$ and $A$, is the set of elements in $B$ but not in $A$.

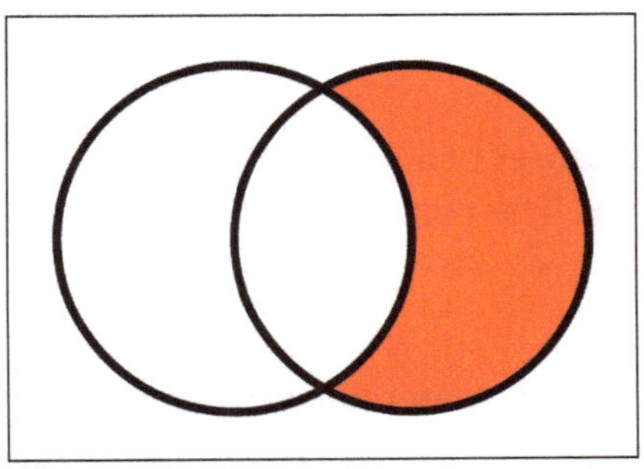

The relative complement of $A$ (left circle) in $B$ (right circle): $B \cap A^c = B \setminus A$.

The relative complement of $A$ in $B$ is denoted $B\setminus A$ according to the ISO 31-11 standard. It is sometimes written $B - A$, but this notation is ambiguous, as in some contexts it can be interpreted as the set of all elements $b - a$, where $b$ is taken from $B$ and $a$ from $A$.

Formally:

$$B \setminus A = \{x \in B \mid x \notin A\}$$

Examples:

- $\{1,2,3\} \setminus \{2,3,4\} = \{1\}$,
- $\{2,3,4\} \setminus \{1,2,3\} = \{4\}$,
- If $\mathbb{R}$ is the set of real numbers and $\mathbb{Q}$. is the set of rational numbers, then $\mathbb{R} \setminus \mathbb{Q}$ is the set of irrational numbers.

## Properties

Let $A$, $B$, and $C$ be three sets. The following identities capture notable properties of relative complements:

- $C \setminus (A \cap B) = (C \setminus A) \cup (C \setminus B)$.
- $C \setminus (A \cup B) = (C \setminus A) \cap (C \setminus B)$.
- $C \setminus (B \setminus A) = (C \cap A) \cup (C \setminus B)$.

with the important special case $C \setminus (C \setminus A) = (C \cap A)$ demonstrating that intersection can be expressed using only the relative complement operation.

- $(B \setminus A) \cap C = (B \cap C) \setminus A = B \cap (C \setminus A)$.
- $(B \setminus A) \cup C = (B \cup C) \setminus (A \setminus C)$.
- $A \setminus A = \varnothing$.
- $\varnothing \setminus A = \varnothing$.
- $A \setminus \varnothing = A$.
- $A \setminus U = \varnothing$.

## Complementary Relation

A binary relation $R$ is defined as a subset of a product of sets $X \times Y$. The complementary relation $\bar{R}$ is the set complement of $R$ in $X \times Y$. The complement of relation $R$ can be written:

$$\bar{R} = (X \times Y) \setminus R.$$

Often $R$ is viewed as a logical matrix with rows representing the elements of $X$, and columns elements of $Y$. The truth of $aRb$ corresponds to 1 in row $a$, column $b$. Producing the complementary relation to $R$ then corresponds to switching all 1s to 0s and 0s to 1s for the logical matrix of the complement.

Together with composition of relations and converse relations, complementary relations and the algebra of sets are the elementary operations of the calculus of relations.

## Cartesian Product

The Cartesian products of sets mean the product of two non-empty sets in an ordered way. Or, in other words, the collection of all ordered pairs obtained by the product of two non-empty sets. An ordered pair means that two elements are taken from each set.

For two non-empty sets (say A & B), the first element of the pair is from one set A and the second element is taken from the second set B. The collection of all such pairs gives us a Cartesian product.

The Cartesian product of two non-empty sets A and B is denoted by A × B. Also, known as the cross-product or the product set of A and B. The ordered pairs (a, b) is such that $a \in A$ and $b \in B$. So, $A \times B = \{(a,b): a \in A, b \in B\}$. For example, Consider two non-empty sets $A = \{a_1, a_2, a_3\}$ and $B = \{b_1, b_2, b_3\}$.

Cartesian product $A \times B = \{(a_1,b_1), (a_1,b_2), (a_1,b_3), (a_2,b_1), (a_2,b_2),(a_2,b_3), (a_3,b_1), (a_3,b_2), (a_3,b_3)\}$.

It is interesting to know that $(a_1,b_1)$ will be different from $(b_1,a_1)$. If either of the two sets is a null set, i.e., either $A = \Phi$ or $B = \Phi$, then, $A \times B = \Phi$ i.e., A × B will also be a null set.

## Number of Ordered Pairs

For two non-empty sets, A and B. If the number of elements of A is h i.e., n(A) = h & that of B is k i.e., n(B) = k, then the number of ordered pairs in Cartesian product will be $n(A \times B) = n(A) \times n(B) = hk$.

## References


- Rosen, Kenneth H. (2012). Discrete Mathematics and Its Applications (7th ed.). New York: mcgraw-Hill. P. 119. ISBN 978-0-07-338309-5
- Introduction-to-sets, sets, maths, guides: toppr.com, Retrieved 31 March, 2019
- Jech, Thomas (2003), Set Theory, Springer Monographs in Mathematics (Third Millennium ed.), Berlin, New York: Springer-Verlag, p. 642, ISBN 978-3-540-44085-7, Zbl 1007.03002
- Fraisse, R. (15 December 2000). Theory of Relations, Volume 145 - 1st Edition (1st ed.). Elsevier. P. 46. ISBN 9780444505422. Retrieved 20 February 2019
- Wagon, Stan; Tomkowicz, Grzegorz (2016). The Banach-Tarski Paradox (Second ed.). ISBN 978-1-107-04259-9.
- Droste, M., & Kuich, W. (2009). Semirings and Formal Power Series. Handbook of Weighted Automata, 3–28. Doi:10.1007/978-3-642-01492-5_1, pp. 7-10
- Cartesian-product-sets, relations-and-functions, +maths, guides: toppr.com, Retrieved 14 July, 2019
- Jech, Thomas (2002), Set Theory, Springer Monographs in Mathematics (3rd millennium ed.), Springer, ISBN 3-540-44085-2

# 3
# Number Theory

Number theory is the branch of pure mathematics that is concerned with the study of integers and integer valued functions. A few of its sub-divisions include algebraic number theory, analytic number theory, additive number theory, combinatorial number theory, etc. All these sub-divisions of number theory have been carefully analyzed in this chapter.

Number theory is a branch of mathematics that studies the properties and relationships of numbers. Specifically, it deals with the natural, or counting numbers, including prime numbers. Number theory is important because the simple sequence of counting numbers from one to infinity conceals many relationships beneath its surface.

## Prime and Composite Numbers

One of the most important distinctions in number theory is between prime and composite numbers. Prime numbers can only be divided evenly (with nothing left over) by 1 and themselves. Prime numbers include 2, 3, 5, 7, 11, 13, 17, and so on to infinity. The number 1 is not considered a prime. All primes are odd numbers except for 2, because any even number can be divided evenly by 2.

A composite number can be divided, or factored, into two or more prime numbers in addition to 1 and itself. Ten is a composite number because it can be divided by 2, 5, 1, and itself. The numbers 2 and 5 are the prime factors of 10. Any whole number that is not a prime is a composite.

One difference between prime and composite numbers is that it takes relatively little time to determine if a number is prime, but far longer to determine the prime factors of a composite number, especially if the composite is very large (100 digits or more). This discrepancy in computation time is important in developing computer security systems.

Prime numbers do not occur in a predictable way. There are sequences of primes which can be partially described in a formula, but sooner or later the formula breaks down. One formula, invented by French philosopher and mathematician Marin Mersenne (1588–1648) is $2^p - 1$, where p is a prime number. Although this formula generates many primes, it also misses many primes. Another formula, invented by Swiss mathematician and physicist Leonhard Euler (1707–1783), generates prime numbers regularly for the series of consecutive numbers from 0 to 15 and then stops. The formula is $x^2 + x + 17$, in which x is any number from 0 to 15.

## Famous Formulas in Number Theory

Number theory is full of famous formulas that illustrate the relationships between whole numbers from 1 to infinity. Some of these formulas are very complicated, but the most famous ones are very simple, for example, the theorem by Fermat below that proves if a number is prime.

## Fermat's Theorem

French mathematician Pierre de Fermat (1601–1665) is one of the most famous number theoreticians in history, but mathematics was only his hobby. Fermat was a judge in France, and he published very little during his life. He did correspond extensively with many leading intellectuals of his day, and his mathematical innovations were presented to these pen pals in his letters.

One of Fermat's many theorems provides a quick way of finding out if a number is prime. Say n is any whole number, and p is any prime number. Raise n to the power of p, and then subtract n from the result. If p is really a prime number, then the result can be divided evenly by p. If anything is left over after the division, then the number p is not prime. A shorter way of putting this formula is this: $n^p - n$ can be divided evenly by p.

Here is a simple illustration of Fermat's theorem. Let n=8 and p=3. If Fermat's theorem is right, then $8^3 - 8$ must be divisible by 3. Multiply 8 by itself three times (8× 8× 8): the product is 512. Subtract 8 from 512: the result is 504. Divide 504 by 3: the result is 168. Fermat's theorem works for any whole numbers that meet the conditions of the formula.

## Gauss and Congruence

German mathematician Karl Friedrich Gauss (1777–1855) made many contributions to pure and applied mathematics. He was born to poor parents in Germany. His high intelligence was noticed early and nurtured by his mother and uncle, but his father never encouraged Gauss in his education.

One of Gauss's most important contributions to number theory involved the invention of the idea of congruence (or agreement) in numbers and the use of what he called "modulos" or small measures or sets of numbers. In effect, his theory of congruence allows people to break up the infinite series of whole numbers into smaller, more manageable chunks of numbers and perform computations upon them. This arrangement makes the everyday arithmetic involved in such things as telling time much easier to program into computers.

Gauss said that if one number is subtracted from another (a - b), and the remainder of the subtraction can be divided by another number, *m,* then a and b are congruent to each other by the number *m*. Gauss's formula is as follows: *a* is congruent to *b* modulo *c*. For example, 720 - 480 = 240. The remainder, 240, can be divided by 60, 20, 10, and other numbers. However, for this purpose, only 60 will be applied. Using Gauss's expression, 720 is congruent to 480 by modulo 60. That is, both 720 and 480 are related to a third number (the remainder after 480 is subtracted from 720), which can be multiplied by 60.

In an abstract sense, this computation is related to such everyday arithmetic functions as telling the time of day on a digital watch. When the watch tells the time, it does not say "240 minutes past noon." It says "4 o'clock" or "4:00." To express the time of day, the digital watch uses several kinds of modulos (or small measures) that have been used for centuries: 60 minutes in an hour, 12 hours in the a.m. or p.m. of a day, and so on. If the watch says it is 4:00 in the afternoon, then, from one frame of reference, it has subtracted 480 minutes from the 720-minute period between 12 noon to midnight. What remains is 240 minutes past noon. That is, 720 - 480 = 240. The remainder, 240, can be divided evenly by the modulo 60 (and by other numbers which will be ignored).

When one tells the time every day, however, people do not use Gauss' terminology. Clocks are already divided into modulos and one simply notes the hour and how many minutes come before or after the hour. The importance of Gauss's congruence theory is that he created the formulas that allowed an immense variety of arithmetic actions to be performed based on different sets of numbers.

## Famous Problems in Number Theory

Number theory is an immensely rich area and it is defined by the important problems that it tries to solve. Sometimes a problem was considered solved, but years later the solution was found to be flawed. One important challenge in number theory has been trying to find a formula that will describe all the prime numbers. To date, that problem has not been solved. Two of the most famous problems in number theory involve Fermat.

## Fermat's Failed Prime Number Formula

Many mathematicians, including Mersenne and Euler, have tried to find a formula that will define all the prime numbers. No one has ever succeeded.

Fermat had one of the most famous failures. He thought that if he squared 2 and then raised the square of 2 to a higher power, which he labeled n (a whole number), then the results would be nothing but primes. His formula looks like this: $_{2}2^{n} + 1$ = [a prime number]. This formula appeared to work until Leonhard Euler proved it wrong. Euler found that if 5 is substituted for n in the formula $_{2}2n + 1$, the resulting number is 4,294,967,297, which can be divided equally by 641 and 6,700,417.

## Fermat's Last Theorem

Fermat wrote his famous Last Theorem in the late 1630s. He said that the equation $x^n + y^n = z^n$ has no solutions in whole numbers if n is greater than 2. Mathematicians have been trying to prove or disprove this theorem for centuries. Princeton University professor Andrew J. Wiles apparently had proved it correct but later flaws were found in his proof. By late 1994, Wiles thought the flaws had been solved. Wiles announced in June 1992 that he had proved Fermat right. In 1998, Christopher Breuil, Brian Conrad, Fred Diamond, and Richard Taylor proved that Wiles had indeed proved Fermat's Last Theorem.

## Applications

Number theory was labeled the Queen of Mathematics by Gauss. For many years, it was thought to be without many practical applications. That situation has changed significantly in the twentieth century with the rise of computers.

Prime and composite numbers play an important role in modern cryptography or coding systems. Huge volumes of confidential information (such as credit card numbers and bank account numbers) and large amounts of money are transferred electronically around the world every day, all of which must kept secret. One of the important applications of number theory is keeping secrets.

Using Fermat's theorem, a computer can quickly compute if a number-even a large number-is prime. However, once a computer finds out that a number is not prime, it then takes a long time to find out what its factors are, especially if the number is a large composite (say 120 digits long). It can take years on a supercomputer to find the prime factors of large composite numbers.

This time gap between finding out if a number is prime and factoring the primes in a composite number is useful to cryptographers. To create a security system, they invent numerical codes for the letters and characters of a message. Then they use an encoding algorithm (a series of steps to solve a problem) to turn a message into a long number. If the message is more than a certain length, say 100 characters, then the cryptography program breaks the message into blocks of 100 characters. Once the message is translated into a number, the program multiplies the number of an encoded message by a certain prime (which could be a 100-digit number) and by a composite number. The composite number is the product of two prime numbers, which have been randomly selected and which must be in both the encoding algorithm of the sender and the decoding algorithm of the receiver. The prime numbers making up the composite number are usually.

- Algorithm — Method for solving a numerical problem consisting of a series of mathematical steps.
- Composite number — A number that can be divided into two or more prime numbers in addition to 1 and itself.
- Congruence — The relationship between two numbers if they have the same remainder when they are divided by a number.
- Cryptography — The study of creating and breaking secret codes.
- Factors — Numbers that are multiplied with other numbers to equal a product. In the multiplication of 2 × 2 = 4, each 2 is a factor. In addition, factors are the numbers that result when numbers are divided. For example, 10 and 10 are factors of 100.
- Modulo — A number by which two other numbers can be divided to give the same remainder.
- Prime numbers — Numbers that can only be divided evenly by 1 and themselves.
- Product — The result of multiplying two or more numbers. Six multiplied by 7 gives the product of 42.
- Set — A set is a collection of things called members or elements of the set. In mathematics, the members of a set will often be numbers.

When the message is transmitted from the sender to the receiver, some of the numbers are made public, but the primes that make up the composite number are kept secret. The decoding algorithm of the authorized person who receives the message only knows them. Anyone who is eavesdropping on the transmission will see many of the numbers, but without the prime numbers from the encoding and decoding programs, it is impossible to decode the message in any reasonable time.

Computer cryptography systems are only one application of number theory. Other formulas of number theory allow computer programs to find out many years in advance what days of the week will fall on what dates of the month, so that people can find out well in advance what day of the week Christmas or the Fourth of July will occur. Many computers have preinstalled internal programs that tell users when they last modified a file down to the second, minute, hour, day of the week, and date of the month. These programs work thanks to the formulas of number theorists.

## Algebraic Number Theory

Algebraic number theory is a branch of number theory that uses the techniques of abstract algebra to study the integers, rational numbers, and their generalizations. Number-theoretic questions are expressed in terms of properties of algebraic objects such as algebraic number fields and their rings of integers, finite fields, and function fields. These properties, such as whether a ring admits unique factorization, the behavior of ideals, and the Galois groups of fields, can resolve questions of primary importance in number theory, like the existence of solutions to Diophantine equations.

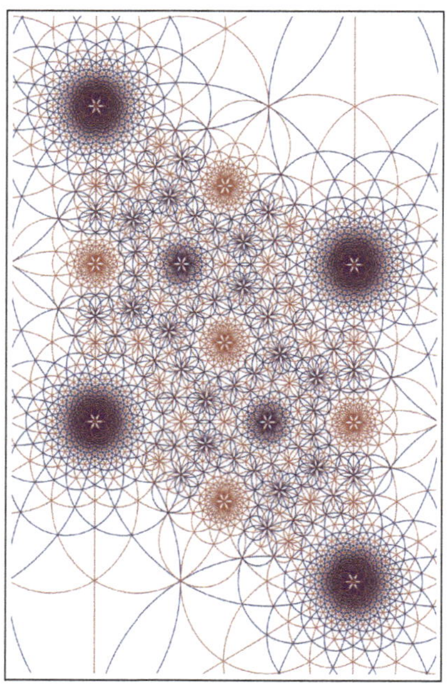

### Basic Notions

### Failure of Unique Factorization

An important property of the ring of integers is that it satisfies the fundamental theorem of arithmetic, that every (positive) integer has a factorization into a product of prime numbers, and this factorization is unique up to the ordering of the factors. This may no longer be true in the ring of integers $O$ of an algebraic number field $K$.

A prime element is an element $p$ of $O$ such that if $p$ divides a product $ab$, then it divides one of the factors $a$ or $b$. This property is closely related to primality in the integers, because any positive integer satisfying this property is either 1 or a prime number. However, it is strictly weaker. For example, −2 is not a prime number because it is negative, but it is a prime element. If factorizations into prime elements are permitted, then, even in the integers, there are alternative factorizations such as:

$$6 = 2 \cdot 3 = (-2) \cdot (-3).$$

In general, if $u$ is a unit, meaning a number with a multiplicative inverse in $O$, and if $p$ is a prime element, then $up$ is also a prime element. Numbers such as $p$ and $up$ are said to be *associate*. In the integers, the primes $p$ and $-p$ are associate, but only one of these is positive. Requiring that prime numbers be positive selects a unique element from among a set of associated prime elements. When $K$ is not the rational numbers, however, there is no analog of positivity. For example, in the Gaussian integers Z[$i$], the numbers $1 + 2i$ and $-2 + i$ are associate because the latter is the product of the former by $i$, but there is no way to single out one as being more canonical than the other. This leads to equations such as:

$$5 = (1 + 2i)(1 - 2i) = (2 + i)(2 - i),$$

which prove that in Z[$i$], it is not true that factorizations are unique up to the order of the factors. For this reason, one adopts the definition of unique factorization used in unique factorization domains (UFDs). In a UFD, the prime elements occurring in a factorization are only expected to be unique up to units and their ordering.

However, even with this weaker definition, many rings of integers in algebraic number fields do not admit unique factorization. There is an algebraic obstruction called the ideal class group. When the ideal class group is trivial, the ring is a UFD. When it is not, there is a distinction between a prime element and an irreducible element. An *irreducible element* $x$ is an element such that if $x = yz$, then either $y$ or $z$ is a unit. These are the elements that cannot be factored any further. Every element in $O$ admits a factorization into irreducible elements, but it may admit more than one. This is because, while all prime elements are irreducible, some irreducible elements may not be prime. For example, consider the ring Z[√-5]. In this ring, the numbers 3, 2 + √-5 and 2 - √-5 are irreducible. This means that the number 9 has two factorizations into irreducible elements:

$$9 = 3^2 = (2+\sqrt{-5})(2-\sqrt{-5}).$$

This equation shows that 3 divides the product (2 + √-5)(2 - √-5) = 9. If 3 were a prime element, then it would divide 2 + √-5 or 2 - √-5, but it does not, because all elements divisible by 3 are of the form $3a + 3b\sqrt{-5}$. Similarly, 2 + √-5 and 2 - √-5 divide the product $3^2$, but neither of these elements divides 3 itself, so neither of them are prime. As there is no sense in which the elements 3, 2 + √-5 and 2 - √-5 can be made equivalent, unique factorization fails in Z[√-5]. Unlike the situation with units, where uniqueness could be repaired by weakening the definition, overcoming this failure requires a new perspective.

## Factorization into Prime Ideals

If $I$ is an ideal in $O$, then there is always a factorization,

$$I = \mathfrak{p}_1^{e_1} \cdots \mathfrak{p}_t^{e_t},$$

where each $\mathfrak{p}_i$ is a prime ideal, and where this expression is unique up to the order of the factors. In particular, this is true if $I$ is the principal ideal generated by a single element. This is the strongest sense in which the ring of integers of a general number field admits unique factorization. In the language of ring theory, it says that rings of integers are Dedekind domains.

When $O$ is a UFD, every prime ideal is generated by a prime element. Otherwise, there are prime ideals which are not generated by prime elements. In Z[√-5], for instance, the ideal (2, 1 + √-5) is a prime ideal which cannot be generated by a single element.

Historically, the idea of factoring ideals into prime ideals was preceded by Ernst Kummer's introduction of ideal numbers. These are numbers lying in an extension field $E$ of $K$. This extension field is now known as the Hilbert class field. By the principal ideal theorem, every prime ideal of $O$ generates a principal ideal of the ring of integers of $E$. A generator of this principal ideal is called an ideal number. Kummer used these as a substitute for the failure of unique factorization in cyclotomic fields. These eventually led Richard Dedekind to introduce a forerunner of ideals and to prove unique factorization of ideals.

An ideal which is prime in the ring of integers in one number field may fail to be prime when extended to a larger number field. Consider, for example, the prime numbers. The corresponding ideals $p$Z are prime ideals of the ring Z. However, when this ideal is extended to the Gaussian integers to get $p$Z[$i$], it may or may not be prime. For example, the factorization 2 = (1 + $i$)(1 − $i$) implies that:

$$2\mathbf{Z}[i] = (1+i)\mathbf{Z}[i] \cdot (1-i)\mathbf{Z}[i] = ((1+i)\mathbf{Z}[i])^2$$

Note that because 1 + $i$ = (1 − $i$) · $i$, the ideals generated by 1 + $i$ and 1 − $i$ are the same. A complete answer to the question of which ideals remain prime in the Gaussian integers is provided by Fermat's theorem on sums of two squares. It implies that for an odd prime number $p$, $p$Z[$i$] is a prime ideal if $p \equiv 3 \pmod 4$ and is not a prime ideal if $p \equiv 1 \pmod 4$. This, together with the observation that the ideal (1 + $i$)Z[$i$] is prime, provides a complete description of the prime ideals in the Gaussian integers. Generalizing this simple result to more general rings of integers is a basic problem in algebraic number theory. Class field theory accomplishes this goal when $K$ is an abelian extension of Q (i.e. a Galois extension with abelian Galois group).

## Ideal Class Group

Unique factorization fails if and only if there are prime ideals that fail to be principal. The object which measures the failure of prime ideals to be principal is called the ideal class group. Defining the ideal class group requires enlarging the set of ideals in a ring of algebraic integers so that they admit a group structure. This is done by generalizing ideals to fractional ideals. A fractional ideal is an additive subgroup $J$ of $K$ which is closed under multiplication by elements of $O$, meaning that $xJ \subseteq J$ if $x \in O$. All ideals of $O$ are also fractional ideals. If $I$ and $J$ are fractional ideals, then the set $IJ$ of all products of an element in $I$ and an element in $J$ is also a fractional ideal. This operation makes the set of non-zero fractional ideals into a group. The group identity is the ideal (1) = $O$, and the inverse of $J$ is a (generalized) ideal quotient, $J^{-1} = (O : J) = \{ x \in K : xJ \subseteq O \}$.

The principal fractional ideals, meaning the ones of the form $Ox$ where $x \in K^\times$, form a subgroup of the group of all non-zero fractional ideals. The quotient of the group of non-zero fractional ideals by this subgroup is the ideal class group. Two fractional ideals $I$ and $J$ represent the same element of the ideal class group if and only if there exists an element $x \in K$ such that $xI = J$. Therefore, the ideal class group makes two fractional ideals equivalent if one is as close to being principal as the other is. The ideal class group is generally denoted Cl $K$, Cl $O$, or Pic $O$ (with the last notation identifying it with the Picard group in algebraic geometry).

The number of elements in the class group is called the class number of $K$. The class number of Q(√-5) is 2. This means that there are only two ideal classes, the class of principal fractional ideals, and the class of a non-principal fractional ideal such as (2, 1 + √-5).

The ideal class group has another description in terms of divisors. These are formal objects which represent possible factorizations of numbers. The divisor group Div $K$ is defined to be the free abelian group generated by the prime ideals of $O$. There is a group homomorphism from $K^\times$, the non-zero elements of $K$ up to multiplication, to Div $K$. Suppose that $x \in K$ satisfies:

$$(x) = \mathfrak{p}_1^{e_1} \cdots \mathfrak{p}_t^{e_t}.$$

Then div $x$ is defined to be the divisor:

$$\operatorname{div} x = \sum_{i=1}^{t} e_i[\mathfrak{p}_i].$$

The kernel of div is the group of units in $O$, while the cokernel is the ideal class group. In the language of homological algebra, this says that there is an exact sequence of abelian groups (written multiplicatively),

$$1 \to O^\times \to K^\times \xrightarrow{\text{div}} \operatorname{Div} K \to \operatorname{Cl} K \to 1.$$

## Real and Complex Embeddings

Some number fields, such as Q(√2), can be specified as subfields of the real numbers. Others, such as Q(√−1), cannot. Abstractly, such a specification corresponds to a field homomorphism $K \to$ R or $K \to$ C. These are called real embeddings and complex embeddings, respectively.

A real quadratic field Q($\sqrt{a}$), with $a \in$ R, $a > 0$, and $a$ not a perfect square, is so-called because it admits two real embeddings but no complex embeddings. These are the field homomorphisms which send $\sqrt{a}$ to $\sqrt{a}$ and to $-\sqrt{a}$, respectively. Dually, an imaginary quadratic field Q($\sqrt{-a}$) admits no real embeddings but admits a conjugate pair of complex embeddings. One of these embeddings sends $\sqrt{-a}$ to $\sqrt{-a}$, while the other sends it to its complex conjugate, $-\sqrt{-a}$.

Conventionally, the number of real embeddings of $K$ is denoted $r_1$, while the number of conjugate pairs of complex embeddings is denoted $r_2$. The signature of $K$ is the pair $(r_1, r_2)$. It is a theorem that $r_1 + 2r_2 = d$, where $d$ is the degree of $K$.

Considering all embeddings at once determines a function:

$$M : K \to \mathbf{R}^{r_1} \oplus \mathbf{C}^{2r_2}.$$

This is called the Minkowski embedding. The subspace of the codomain fixed by complex conjugation is a real vector space of dimension $d$ called Minkowski space. Because the Minkowski embedding is defined by field homomorphisms, multiplication of elements of $K$ by an element $x \in K$ corresponds to multiplication by a diagonal matrix in the Minkowski embedding. The dot product on Minkowski space corresponds to the trace form $\langle x, y\rangle = \mathrm{Tr}(xy)$

The image of $O$ under the Minkowski embedding is a $d$-dimensional lattice. If $B$ is a basis for this lattice, then det $B^TB$ is the discriminant of $O$. The discriminant is denoted $\Delta$ or $D$. The covolume of the image of $O$ is $\sqrt{|\Delta|}$

## Places

Real and complex embeddings can be put on the same footing as prime ideals by adopting a perspective based on valuations. Consider, for example, the integers. In addition to the usual absolute value function $|\cdot| : Q \to R$, there are p-adic absolute value functions $|\cdot|_p : Q \to R$, defined for each prime number $p$, which measure divisibility by $p$. Ostrowski's theorem states that these are all possible absolute value functions on Q (up to equivalence). Therefore, absolute values are a common language to describe both the real embedding of Q and the prime numbers.

A place of an algebraic number field is an equivalence class of absolute value functions on $K$. There are two types of places. There is a p-adic absolute value for each prime ideal p of $O$, and, like the $p$-adic absolute values, it measures divisibility. These are called finite places. The other type of place is specified using a real or complex embedding of $K$ and the standard absolute value function on R or C. These are infinite places. Because absolute values are unable to distinguish between a complex embedding and its conjugate, a complex embedding and its conjugate determine the same place. Therefore, there are $r_1$ real places and $r_2$ complex places. Because places encompass the primes, places are sometimes referred to as primes. When this is done, finite places are called finite primes and infinite places are called infinite primes. If $v$ is a valuation corresponding to an absolute value, then one frequently writes $v \mid \infty$ to mean that $v$ is an infinite place and $v \nmid \infty$ to mean that it is a finite place.

Considering all the places of the field together produces the adele ring of the number field. The adele ring allows one to simultaneously track all the data available using absolute values. This produces significant advantages in situations where the behavior at one place can affect the behavior at other places, as in the Artin reciprocity law.

## Units

The integers have only two units, 1 and −1. Other rings of integers may admit more units. The Gaussian integers have four units, the previous two as well as $\pm i$. The Eisenstein integers Z[exp($2\pi i$ / 3)] have six units. The integers in real quadratic number fields have infinitely many units. For example, in Z[√3], every power of 2 + √3 is a unit, and all these powers are distinct.

In general, the group of units of $O$, denoted $O^\times$, is a finitely generated abelian group. The fundamental theorem of finitely generated abelian groups therefore implies that it is a direct sum of a torsion part and a free part. Reinterpreting this in the context of a number field, the torsion part consists of the roots of unity that lie in $O$. This group is cyclic. The free part is described by Dirichlet's unit theorem. This theorem says that rank of the free part is $r_1 + r_2 - 1$. Thus, for example, the only fields for which the rank of the free part is zero are Q and the imaginary quadratic fields. A more precise statement giving the structure of $O^\times \otimes_Z Q$ as a Galois module for the Galois group of $K/Q$ is also possible.

The free part of the unit group can be studied using the infinite places of $K$. Consider the function,

$$L : K^\times \to \mathbf{R}^{r_1+r_2}$$

defined by,

$$L(x) = (\log |x|_v)_v,$$

where $v$ varies over the infinite places of $K$ and $|\cdot|_v$ is the absolute value associated with $v$. The function $L$ is a homomorphism from $K^\times$ to a real vector space. It can be shown that the image of $O^\times$ is a lattice that spans the hyperplane defined by $x_1 + \cdots + x_{r_1+r_2} = 0$. The covolume of this lattice is the regulator of the number field. One of the simplifications made possible by working with the adele ring is that there is a single object, the idele class group, that describes both the quotient by this lattice and the ideal class group.

## Zeta Function

The Dedekind zeta function of a number field, analogous to the Riemann zeta function is an analytic object which describes the behavior of prime ideals in $K$. When $K$ is an abelian extension of Q, Dedekind zeta functions are products of Dirichlet L-functions, with there being one factor for each Dirichlet character. The trivial character corresponds to the Riemann zeta function. When $K$ is a Galois extension, the Dedekind zeta function is the Artin L-function of the regular representation of the Galois group of $K$, and it has a factorization in terms of irreducible Artin representations of the Galois group.

The zeta function is related to the other invariants by the class number formula.

## Local Fields

Completing a number field $K$ at a place $w$ gives a complete field. If the valuation is archimedean, one gets R or C, if it is non-archimedean and lies over a prime $p$ of the rationals, one gets a finite extension $K_w/Q_p$: a complete, discrete valued field with finite residue field. This process simplifies the arithmetic of the field and allows the local study of problems. For example, the Kronecker–Weber theorem can be deduced easily from the analogous local statement. The philosophy behind the study of local fields is largely motivated by geometric methods. In algebraic geometry, it is common to study varieties locally at a point by localizing to a maximal ideal. Global information can then be recovered by gluing together local data. This spirit is adopted in algebraic number theory. Given a prime in the ring of algebraic integers in a number field, it is desirable to study the field locally at that prime. Therefore, one localizes the ring of algebraic integers to that prime and then completes the fraction field much in the spirit of geometry.

## Major Results

### Finiteness of the Class Group

One of the classical results in algebraic number theory is that the ideal class group of an algebraic number field $K$ is finite. The order of the class group is called the class number, and is often denoted by the letter $h$.

### Dirichlet's Unit Theorem

Dirichlet's unit theorem provides a description of the structure of the multiplicative group of units $O^{\times}$ of the ring of integers $O$. Specifically, it states that $O^{\times}$ is isomorphic to $G \times Z^r$, where $G$ is the finite cyclic group consisting of all the roots of unity in $O$, and $r = r_1 + r_2 - 1$ (where $r_1$ (respectively, $r_2$) denotes the number of real embeddings (respectively, pairs of conjugate non-real embeddings) of $K$). In other words, $O^{\times}$ is a finitely generated abelian group of rank $r_1 + r_2 - 1$ whose torsion consists of the roots of unity in $O$.

### Reciprocity Laws

In terms of the Legendre symbol, the law of quadratic reciprocity for positive odd primes states:

$$\left(\frac{p}{q}\right)\left(\frac{q}{p}\right)=(-1)^{\frac{p-1}{2}\frac{q-1}{2}}.$$

A reciprocity law is a generalization of the law of quadratic reciprocity.

There are several different ways to express reciprocity laws. The early reciprocity laws found in the 19th century were usually expressed in terms of a power residue symbol $(p/q)$ generalizing the quadratic reciprocity symbol, that describes when a prime number is an $n$th power residue modulo another prime, and gave a relation between $(p/q)$ and $(q/p)$. Hilbert reformulated the reciprocity laws as saying that a product over $p$ of Hilbert symbols $(a,b/p)$, taking values in roots of unity, is equal to 1. Artin's reformulated reciprocity law states that the Artin symbol from ideals (or ideles) to elements of a Galois group is trivial on a certain subgroup. Several more recent generalizations express reciprocity laws using cohomology of groups or representations of adelic groups or algebraic K-groups, and their relationship with the original quadratic reciprocity law can be hard to see.

### Class Number Formula

The class number formula relates many important invariants of a number field to a special value of its Dedekind zeta function.

# Analytic Number Theory

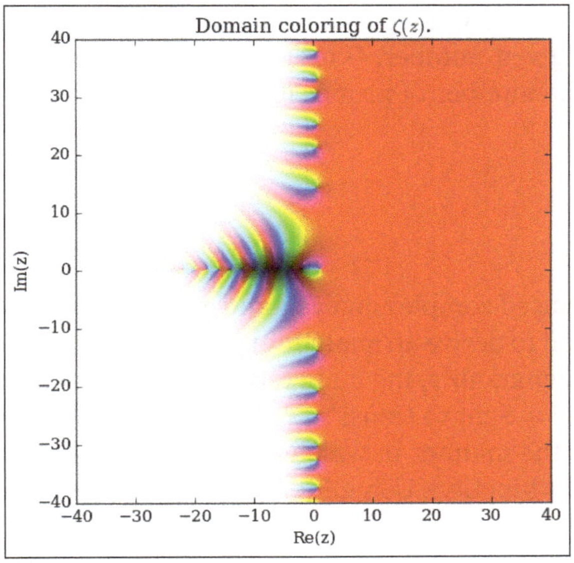


Riemann zeta function $\zeta(s)$ in the complex plane. The color of a point $s$ encodes the value of $\zeta(s)$: colors close to black denote values close to zero, while hue encodes the value's argument.

In mathematics, analytic number theory is a branch of number theory that uses methods from mathematical analysis to solve problems about the integers. It is often said to have begun with Peter Gustav Lejeune Dirichlet's 1837 introduction of Dirichlet $L$-functions to give the first proof of Dirichlet's theorem on arithmetic progressions. It is well known for its results on prime numbers (involving the Prime Number Theorem and Riemann zeta function) and additive number theory (such as the Goldbach conjecture and Waring's problem).

## Branches of Analytic Number Theory

Analytic number theory can be split up into two major parts, divided more by the type of problems they attempt to solve than fundamental differences in technique.

- Multiplicative number theory deals with the distribution of the prime numbers, such as estimating the number of primes in an interval, and includes the prime number theorem and Dirichlet's theorem on primes in arithmetic progressions.
- Additive number theory is concerned with the additive structure of the integers, such as Goldbach's conjecture that every even number greater than 2 is the sum of two primes. One of the main results in additive number theory is the solution to Waring's problem.

## Problems and Results

Theorems and results within analytic number theory tend not to be exact structural results about the integers, for which algebraic and geometrical tools are more appropriate. Instead, they give approximate bounds and estimates for various number theoretical functions, as the following examples illustrate.

## Multiplicative Number Theory

Euclid showed that there are infinitely many prime numbers. An important question is to determine the asymptotic distribution of the prime numbers; that is, a rough description of how many primes are smaller than a given number. Gauss, amongst others, after computing a large list of primes, conjectured that the number of primes less than or equal to a large number $N$ is close to the value of the integral:

$$\int_2^N \frac{1}{\log t} dt.$$

In 1859 Bernhard Riemann used complex analysis and a special meromorphic function now known as the Riemann zeta function to derive an analytic expression for the number of primes less than or equal to a real number $x$. Remarkably, the main term in Riemann's formula was exactly the above integral, lending substantial weight to Gauss's conjecture. Riemann found that the error terms in this expression, and hence the manner in which the primes are distributed, are closely related to the complex zeros of the zeta function. Using Riemann's ideas and by getting more information on the zeros of the zeta function, Jacques Hadamard and Charles Jean de la Vallée-Poussin managed to complete the proof of Gauss's conjecture. In particular, they proved that if,

$$\pi(x) = (\text{number of primes } \leq x),$$

then:

$$\lim_{x\to\infty} \frac{\pi(x)}{x/\log x} = 1.$$

This remarkable result is what is now known as the *prime number theorem*. It is a central result in analytic number theory. Loosely speaking, it states that given a large number $N$, the number of primes less than or equal to $N$ is about $N/\log(N)$.

More generally, the same question can be asked about the number of primes in any arithmetic progression $a+nq$ for any integer $n$. In one of the first applications of analytic techniques to number theory, Dirichlet proved that any arithmetic progression with $a$ and $q$ coprime contains infinitely many primes. The prime number theorem can be generalised to this problem; letting,

$$\pi(x,a,q) = (\text{number of primes } \leq x \text{ such that } p \text{ is in the arithmetic progression } a+nq, n \in \mathbf{Z}),$$

then if $a$ and $q$ are coprime,

$$\lim_{x\to\infty} \frac{\pi(x,a,q)\phi(q)}{x/\log x} = 1.$$

There are also many deep and wide-ranging conjectures in number theory whose proofs seem too difficult for current techniques, such as the twin prime conjecture which asks whether there are infinitely many primes $p$ such that $p + 2$ is prime. On the assumption of the Elliott–Halberstam conjecture it has been proven recently that there are infinitely many primes $p$ such that $p + k$ is prime for some positive even $k$ at most 12. Also, it has been proven unconditionally (i.e. not depending on unproven conjectures) that there are infinitely many primes $p$ such that $p + k$ is prime for some positive even $k$ at most 246.

## Additive Number Theory

One of the most important problems in additive number theory is Waring's problem, which asks whether it is possible, for any $k \geq 2$, to write any positive integer as the sum of a bounded number of $k$th powers,

$$n = x_1^k + \cdots + x_\ell^k.$$

The case for squares, $k = 2$, was answered by Lagrange in 1770, who proved that every positive integer is the sum of at most four squares. The general case was proved by Hilbert in 1909, using algebraic techniques which gave no explicit bounds. An important breakthrough was the application of analytic tools to the problem by Hardy and Littlewood. These techniques are known as the circle method, and give explicit upper bounds for the function $G(k)$, the smallest number of $k$th powers needed, such as Vinogradov's bound,

$$G(k) \leq k(3\log k + 11).$$

## Diophantine Problems

Diophantine problems are concerned with integer solutions to polynomial equations: one may study the distribution of solutions, that is, counting solutions according to some measure of "size" or *height*.

An important example is the Gauss circle problem, which asks for integers points ($x$ $y$) which satisfy,

$$x^2 + y^2 \leq r^2.$$

In geometrical terms, given a circle centered about the origin in the plane with radius $r$, the problem asks how many integer lattice points lie on or inside the circle. It is not hard to prove that the answer is $\pi r^2 + E(r)$, where $E(r)/r^2 \to 0$ as $r \to \infty$ Again, the difficult part and a great achievement of analytic number theory is obtaining specific upper bounds on the error term $E(r)$.

It was shown by Gauss that $E(r) = O(r)$. In general, an $O(r)$ error term would be possible with the unit circle (or, more properly, the closed unit disk) replaced by the dilates of any bounded planar region with piecewise smooth boundary. Furthermore, replacing the unit circle by the unit square, the error term for the general problem can be as large as a linear function of $r$. Therefore, getting an error bound of the form $O(r^\delta)$ for some $\delta < 1$ in the case of the circle is a significant improvement. The first to attain this was Sierpiński in 1906, who showed $E(r) = O(r^{2/3})$. In 1915, Hardy and Landau each showed that one does *not* have $E(r) = O(r^{1/2})$. Since then the goal has been to show that for each fixed $\epsilon > 0$ there exists a real number $C(\dot{o})$ such that $E(r) \leq C(\epsilon) r^{1/2+\epsilon}$.

In 2000 Huxley showed that $E(r) = O(r^{131/208})$, which is the best published result.

## Methods of Analytic Number Theory

### Dirichlet Series

One of the most useful tools in multiplicative number theory are Dirichlet series, which are functions of a complex variable defined by an infinite series of the form:

$$f(s)=\sum_{n=1}^{\infty}a_n n^{-s}.$$

Depending on the choice of coefficients $a_n$ this series may converge everywhere, nowhere, or on some half plane. In many cases, even where the series does not converge everywhere, the holomorphic function it defines may be analytically continued to a meromorphic function on the entire complex plane. The utility of functions like this in multiplicative problems can be seen in the formal identity,

$$\left(\sum_{n=1}^{\infty}a_n n^{-s}\right)\left(\sum_{n=1}^{\infty}b_n n^{-s}\right)=\sum_{n=1}^{\infty}\left(\sum_{k\ell=n}a_k b_\ell\right)n^{-s};$$

hence the coefficients of the product of two Dirichlet series are the multiplicative convolutions of the original coefficients. Furthermore, techniques such as partial summation and Tauberian theorems can be used to get information about the coefficients from analytic information about the Dirichlet series. Thus a common method for estimating a multiplicative function is to express it as a Dirichlet series (or a product of simpler Dirichlet series using convolution identities), examine this series as a complex function and then convert this analytic information back into information about the original function.

### Riemann Zeta Function

Euler showed that the fundamental theorem of arithmetic implies (at least formally) the Euler product,

$$\sum_{n=1}^{\infty}\frac{1}{n^s}=\prod_{p}^{\infty}\frac{1}{1-p^{-s}} \text{ for } s>1 \ (p \text{ is prime number})$$

Euler's proof of the infinity of prime numbers makes use of the divergence of the term at the left hand side for $s = 1$ (the so-called harmonic series), a purely analytic result. Euler was also the first to use analytical arguments for the purpose of studying properties of integers, specifically by constructing generating power series. This was the beginning of analytic number theory.

Later, Riemann considered this function for complex values of $s$ and showed that this function can be extended to a meromorphic function on the entire plane with a simple pole at $s = 1$. This function is now known as the Riemann Zeta function and is denoted by $\zeta(s)$. There is a plethora of literature on this function and the function is a special case of the more general Dirichlet L-functions.

Analytic number theorists are often interested in the error of approximations such as the prime number theorem. In this case, the error is smaller than $x/\log x$. Riemann's formula for $\pi(x)$ shows that the error term in this approximation can be expressed in terms of the zeros of the zeta function. In his 1859 paper, Riemann conjectured that all the "non-trivial" zeros of $\zeta$ lie on the line $\Re(s) = 1/2$ but never provided a proof of this statement. This famous and long-standing conjecture is known as the *Riemann Hypothesis* and has many deep implications in number theory; in fact, many important theorems have been proved under the assumption that the hypothesis is true. For example, under the assumption of the Riemann Hypothesis, the error term in the prime number theorem is $O(x^{1/2+\varepsilon})$.

In the early 20th century G. H. Hardy and Littlewood proved many results about the zeta function in an attempt to prove the Riemann Hypothesis. In fact, in 1914, Hardy proved that there were infinitely many zeros of the zeta function on the critical line:

$$\Re(z) = 1/2.$$

This led to several theorems describing the density of the zeros on the critical line.

# Computational Number Theory

In mathematics and computer science, computational number theory, also known as algorithmic number theory, is the study of algorithms for performing number theoretic computations. The aim of the computational number theory is to study the most relevant algorithm from the number theory.

## Algorithms

There are many algorithms in the computational number theory. The following are some of the algorithms and their complexities.

## Multiplication

The lower bound for the computational complexity of multiplication has been a focus since the first computer appeared. The standard multiplication algorithm has complexity $O(n^2)$, but the time complexity has been long conjectured to be O(n log(n)), which was proved in 2019.

## Exponential

As the multiplication, exponential has been an aim for the number theory. If we try to compute an exponential operation with successive multiplication, we get a complexity as same as the multiplication with the standard algorithm ($O(n^2)$). To improve this, we can use the repeated multiplication and reduction algorithm. With this, we reduce the complexity to $O(M(n)\ 2^k)$. Other algorithms are Exponentiation by squaring or Montgomery reduction that reduce the complexity to O(M(n) k).

## Primality Test

A very important part of the computational number theory is to know if a number is prime. This can be difficult because it requires a large amount of computer power to prove primality in a large number.

An easier algorithm to help identify if a number is prime is a successive division by all of its lower numbers. If we depart from that, the standard division takes a complexity of $O(n^2)$. If we want to know if a number $n$ is prime, then the global complexity is $O(n \times n^2)$. To improve this, we can go to AKS primality test, and if we assume Agrawal's conjecture, the complexity is $O((\log n)^3)$; otherwise it takes $O((\log n)^6)$.

### Integer Factorials

The factorial is the product of all the positive integer numbers from 1 to the number $n$. With successive multiplication, the factorial can be found using a complexity of $O(n \times n^{2)})$. With the Bottom-up multiplication algorithm, this complexity can be reduced to $O(M(m^2) \log m)$.

### Greatest Common Divisor

The greatest common divisor is a main scope of the computational number theory. One method to calculate this is by comparing all the divisors between two numbers (however this is inefficient). The computational number theory has developed algorithms to improve this, such as the Euclidean algorithm, which can compute the GCD with complexity of $O(n^2)$. There is also the Binary GCD which can solve this problem with the same complexity. There are other algorithms that go further, such as the Stehlé–Zimmermann algorithm, which can get the GCD with the complexity of $O(M(n) \log n)$.

## Additive Number Theory

In number theory, the specialty additive number theory studies subsets of integers and their behavior under addition. More abstractly, the field of "additive number theory" includes the study of abelian groups and commutative semigroups with an operation of addition. Additive number theory has close ties to combinatorial number theory and the geometry of numbers. Two principal objects of study are the sumset of two subsets $A$ and $B$ of elements from an abelian group $G$,

$$A+B=\{a+b : a \in A, b \in B\}$$

and the h-fold sumset of $A$,

$$hA = \underbrace{A+\cdots+A}_{h}$$

The field is principally devoted to consideration of *direct problems* over (typically) the integers, that is, determining the structure of $hA$ from the structure of $A$: for example, determining which elements can be represented as a sum from $hA$, where, $A$ is a fixed subset. Two classical problems of this type are the Goldbach conjecture (which is the conjecture that $2P$ contains all even numbers greater than two, where $P$ is the set of primes) and Waring's problem (which asks how large must $h$ be to guarantee that $hA_k$ contains all positive integers, where,

$$A_k = \{0^k, 1^k, 2^k, 3^k, \ldots\}$$

is the set of k-th powers). Many of these problems are studied using the tools from the Hardy-Littlewood circle method and from sieve methods. For example, Vinogradov proved that every sufficiently large odd number is the sum of three primes, and so every sufficiently large even integer is the sum of four primes. Hilbert proved that, for every integer $k > 1$, every nonnegative integer is the sum of a bounded number of $k$-th powers. In general, a set $A$ of nonnegative integers is called a basis of order $h$ if $hA$ contains all positive integers, and it is called an asymptotic basis if $hA$ contains all sufficiently large integers. Much current research in this area concerns properties of general asymptotic bases of finite order. For example, a set $A$ is called a minimal asymptotic basis of order $h$ if $A$ is an asymptotic basis of order h but no proper subset of $A$ is an asymptotic basis of order $h$. It has been proved that minimal asymptotic bases of order $h$ exist for all $h$, and that there also exist asymptotic bases of order $h$ that contain no minimal asymptotic bases of order $h$. Another question to be considered is how small can the number of representations of $n$ as a sum of $h$ elements in an asymptotic basis can be. This is the content of the Erdős–Turán conjecture on additive bases.

### Additive Combinatorics

The second is principally devoted to consideration of *inverse problems*, often over more general groups than just the integers, that is, given some information about the sumset $A+B$, the aim is find information about the structure of the individual sets $A$ and $B$. (A more recent name sometimes associated to this sub-division is additive combinatorics.) Unlike problems related to classical bases, this sub-area often deals with finite subsets rather than infinite ones. A typical question is what is the structure of a pair of subsets whose sumset has small cardinality (in relation to $|A|$ and $|B|$). In the case of the integers, the classical Freiman's theorem provides a potent partial answer to this question in terms of multi-dimensional arithmetic progressions. Another typical problem is simply to find a lower bound for $|A+B|$ in terms of $|A|$ and $|B|$ (this can be viewed as an inverse problem with the given information for $A+B$ being that $|A+B|$ is sufficiently small and the structural conclusion then being that either $A$ or $B$ is the empty set; such problems are often considered direct problems as well). Examples of this type include the Erdős–Heilbronn Conjecture (for a restricted sumset) and the Cauchy–Davenport Theorem. The methods used for tackling such questions draw from across the spectrum of mathematics, including combinatorics, ergodic theory, analysis, graph theory, group theory, and linear algebraic and polynomial methods.

## Probabilistic Number Theory

In mathematics, Probabilistic number theory is a subfield of number theory, which explicitly uses probability to answer questions of number theory. One basic idea underlying it is that different prime numbers are, in some serious sense, like independent random variables. This however is not an idea that has a unique useful formal expression.

The founders of the theory were Paul Erdős, Aurel Wintner and Mark Kac during the 1930s, one of the periods of investigation in analytic number theory. The Erdős–Wintner theorem and the Erdős–Kac theorem on additive functions were foundational results.

# Combinatorial Number Theory

Combinatorial number theory can be described as an area of mathematics which is a cross between combinatorics and number theory. More precisely, the area concerns structures of integers (or similar sets), with some number theoretic properties, which can be studied mainly by combinatorial means, rather than, for example, by algebraic methods.

Problems and results in combinatorial number theory really cannot be rigorously classified, but here is an effort at exposition of a few broad categories. These represent just some of the major areas of combinatorial number theory and are by no means intended to represent the field completely.

## Divisibility Issues

One basic area of study is related to the relation of divisibility between positive integers and its interplay with addition and ordering. Interpreting a number in terms of its prime factorization often simplifies questions of divisibility: a number is divisible by another if and only if each prime has an exponent in the first number at least as large as its exponent in the second; this formulation facilitates calculation of greatest common divisors and least common multiples. However, this interpretation of divisibility is far from trivializing many problems and results, and a variety of other techniques prove essential as well in the study of the seemingly simple notions of divisibility, relative primality, and so forth. Because the set of positive integers, together with the relation of divisibility, constitutes one of the basic examples of partially ordered sets, it is of considerable combinatorial importance.

## Partitions of Sets of Integers

Another important area of combinatorial number theory is the study of what happens when the positive integers, or some finite subset thereof, are distributed among a bunch of smaller sets, often phrased in terms of "coloring" the integers. It is of particular interest to define a "largeness" or "density" property for sets of positive integers, and to show that, when the positive integers are divided into finitely many subsets, some subset is "large." For example, a result of Schur states that, for every positive integer k, there is some n with the following property: when the integers from 1 through n are divided into k subsets, some subset contains three distinct numbers such that one is the sum of the other two. Another significant example is Van der Waerden's theorem that, if the set of all positive integers is divided into finitely many subsets, there is some subset which contains arbitrarily long arithmetic progressions. In addition to finding properties that one subset must have, one also studies relations between the subsets. This study is particularly useful when one restricts one's attention to partitions of the integers into particular types of subsets, such as arithmetic progressions. A variety of combinatorial and number theoretic techniques like - construction, induction, contradiction, direct proofs - are used in this area.

## Additive Problems

One of the largest areas of combinatorial number theory - and one of the broadest, as it connects not only with combinatorics but also analysis and algebra - is additive number theory: the study of what happens when sets of integers are added together. One of the most famous unsolved problems in number theory is an additive one: the Goldbach conjecture, which says that every even number greater than 4 is the sum of two odd primes. Many combinatorial and algebraic techniques prove extremely valuable here; most notable is the pigeonhole principle, but other methods of approaching problems include classifying sets of numbers to be added, the use of minimal and maximal elements, counting sets in multiple ways, and the representation of a sum of numbers as a difference of two other sums.

## Partitions

The deepest and most serious area of research in combinatorial number theory is concerned with partitions of a positive integer. A partition is a representation of an integer as a sum of other positive integers, called the parts. Two partitions with the same parts in a different order are considered the same. For example, 4 has 5 partitions: 4, 3 + 1, 2 + 2, 2 + 1 + 1, and 1 + 1 + 1 + 1. Many colorful results are concerned with the number of partitions of a positive integer or the number of partitions satisfying a particular condition. A famous result of Ramanujan (who did extensive work on partitions) asserts that the number of partitions of a number of the form 5k + 4 is always divisible by 5. Also important is Euler's "pentagonal number theorem," which provides a recurrence relation for calculating the number p(n) of partitions of n: it equals the sum of $(-1)^{m+1}$ $\left(p\left(n-\frac{m(3m-1)}{2}\right)+p\left(n-\frac{m(3m+1)}{2}\right)\right)$ over all positive integers m, where we take p(0) = 1 and p(n) = 0 if n < 0. Partitions played a crucial role in the recent proof of the alternating sign matrix conjecture.

Unlike many other areas of combinatorial number theory, there are a few specific techniques which are of considerable importance in examining partitions. One is the use of Ferrers diagrams, a method of representing partitions graphically; also useful is the technique of bijection, where one finds a function that turns one type of partition into another type of partition and thereby shows that the number of partitions of the first type equals the number of partitions of the second type. By far the most significant tool, however, is the use of the generating function. A generating function is a power series (or "infinite polynomial") in which the coefficients represent a significant sequence of numbers. By doing algebraic operations on entire series, one can obtain information about the individual coefficients. Sequences where the nth coefficients represents information about partitions of n happen to be particularly amenable to this kind of manipulation.

## References


- Buhler, J.P.; P., Stevenhagen, eds. (2008). Algorithmic Number Theory: Lattices, Number Fields, Curves and Cryptography. MSRI Publications. 44. Cambridge University Press. ISBN 978-0-521-20833-8. Zbl 1154.11002
- Number-theory, mathematics, science-and-technology: encyclopedia.com, Retrieved 17 May, 2019
- Davenport, Harold (2000), Multiplicative number theory, Graduate Texts in Mathematics, 74 (3rd revised ed.), New York: Springer-Verlag, ISBN 978-0-387-95097-6, MR 1790423
- Samuel S. Wagstaff, Jr. (2013). The Joy of Factoring. Providence, RI: American Mathematical Society. ISBN 978-1-4704-1048-3

# 4
# Combinatorics

Combinatorics is the field of mathematics that deals with the problems of selection, configuration and operation within a discrete system. Some of its important aspects include combinatorial modeling, Pascal's identity, permutation, etc. This chapter closely examines these key aspects of combinatorics to provide an extensive understanding of the subject.

Combinatorics also called combinatorial mathematics is the field of mathematics concerned with problems of selection, arrangement, and operation within a finite or discrete system. Included is the closely related area of combinatorial geometry.

One of the basic problems of combinatorics is to determine the number of possible configurations (*e.g.*, graphs, designs, arrays) of a given type. Even when the rules specifying the configuration are relatively simple, enumeration may sometimes present formidable difficulties. The mathematician may have to be content with finding an approximate answer or at least a good lower and upper bound.

In mathematics, generally, an entity is said to "exist" if a mathematical example satisfies the abstract properties that define the entity. In this sense it may not be apparent that even a single configuration with certain specified properties exists. This situation gives rise to problems of existence and construction. There is again an important class of theorems that guarantee the existence of certain choices under appropriate hypotheses. Besides their intrinsic interest, these theorems may be used as existence theorems in various combinatorial problems.

Finally, there are problems of optimization. As an example, a function $f$, the economic function, assigns the numerical value $f(x)$ to any configuration $x$ with certain specified properties. In this case the problem is to choose a configuration $x_0$ that minimizes $f(x)$ or makes it $\varepsilon$ = minimal—that is, for any number $\varepsilon > 0$, $f(x_0) f(x) + \varepsilon$, for all configurations $x$, with the specified properties.

## Combinatorial Number System

In mathematics, and in particular in combinatorics, the combinatorial number system of degree $k$ (for some positive integer $k$), also referred to as combinadics, is a correspondence between natural numbers (taken to include 0) $N$ and $k$-combinations, represented as strictly decreasing sequences

$c_k > \ldots > c_2 > c_1 \geq 0$. Since the latter are strings of numbers, one can view this as a kind of numeral system for representing $N$, although the main utility is representing a $k$-combination by $N$ rather than the other way around. Distinct numbers correspond to distinct $k$-combinations, and produce them in lexicographic order; moreover the numbers less than $\binom{n}{k}$ correspond to all $k$-combinations of $\{0, 1, \ldots, n-1\}$. The correspondence does not depend on the size $n$ of the set that the $k$-combinations are taken from, so it can be interpreted as a map from N to the $k$-combinations taken from N; in this view the correspondence is a bijection.

The number $N$ corresponding to $(c_k,\ldots,c_2,c_1)$ is given by:

$$N = \binom{c_k}{k} + \cdots + \binom{c_2}{2} + \binom{c_1}{1}.$$

The fact that a unique sequence so corresponds to any number $N$ was observed by D. H. Lehmer. Indeed a greedy algorithm finds the $k$-combination corresponding to $N$: take $c_k$ maximal with $\binom{c_k}{k} \leq N$, then take $c_{k-1}$ maximal with $\binom{c_{k-1}}{k-1} \leq N - \binom{c_k}{k}$, and so forth. Finding the number $N$, using the formula above, from the $k$-combination $(c_k,\ldots,c_2,c_1)$ is also known as "ranking", and the opposite operation (given by the greedy algorithm) as "unranking"; these operations are known by those names in most computer algebra systems, and in computational mathematics.

The originally used term "combinatorial representation of integers" is shortened to "combinatorial number system" by Knuth, who also gives a much older reference; the term "combinadic" is introduced by James McCaffrey (without reference to previous terminology or work).

Unlike the factorial number system, the combinatorial number system of degree $k$ is not a mixed radix system: the part $\binom{c_i}{i}$ of the number $N$ represented by a "digit" $c_i$ is not obtained from it by simply multiplying by a place value.

The main application of the combinatorial number system is that it allows rapid computation of the $k$-combination that is at a given position in the lexicographic ordering, without having to explicitly list the $k$-combinations preceding it; this allows for instance random generation of $k$-combinations of a given set. Enumeration of $k$-combinations has many applications, among which software testing, sampling, quality control, and the analysis of lottery games.

## Ordering Combinations

A $k$-combination of a set $S$ is a subset of $S$ with $k$ (distinct) elements. The main purpose of the combinatorial number system is to provide a representation, each by a single number, of all $\binom{n}{k}$ possible $k$-combinations of a set $S$ of $n$ elements. Choosing, for any $n$, $\{0, 1, \ldots, n-1\}$ as such a set, it can be arranged that the representation of a given $k$-combination $C$ is independent of the value of $n$ (although $n$ must of course be sufficiently large); in other words considering $C$ as a subset of a larger set by increasing $n$ will not change the number that represents $C$. Thus for the combinatorial number system one just considers $C$ as a $k$-combination of the set N of all natural numbers, without explicitly mentioning $n$.

In order to ensure that the numbers representing the $k$-combinations of {0, 1, ..., $n$ – 1} are less than those representing $k$-combinations not contained in {0, 1, ..., $n$ – 1}, the $k$-combinations must be ordered in such a way that their largest elements are compared first. The most natural ordering that has this property is lexicographic ordering of the *decreasing* sequence of their elements. So comparing the 5-combinations $C$ = {0,3,4,6,9} and $C'$ = {0,1,3,7,9}, one has that $C$ comes before $C'$, since they have the same largest part 9, but the next largest part 6 of $C$ is less than the next largest part 7 of $C'$; the sequences compared lexicographically are (9,6,4,3,0) and (9,7,3,1,0). Another way to describe this ordering is view combinations as describing the $k$ raised bits in the binary representation of a number, so that $C = \{c_1,...,c_k\}$ describes the number,

$$2^{c_1} + 2^{c_2} + \cdots + 2^{c_k}$$

This associates distinct numbers to all finite sets of natural numbers; then comparison of $k$-combinations can be done by comparing the associated binary numbers. In the example $C$ and $C'$ correspond to numbers $1001011001_2 = 601_{10}$ and $1010001011_2 = 651_{10}$, which again shows that $C$ comes before $C'$. This number is not however the one one wants to represent the $k$-combination with, since many binary numbers have a number of raised bits different from $k$; one wants to find the relative position of $C$ in the ordered list of (only) $k$-combinations.

## Place of a Combination in the Ordering

The number associated in the combinatorial number system of degree $k$ to a $k$-combination $C$ is the number of $k$-combinations strictly less than $C$ in the given ordering. This number can be computed from $C = \{ c_k, ..., c_2, c_1 \}$ with $c_k > ... > c_2 > c_1$ as follows. From the definition of the ordering it follows that for each $k$-combination $S$ strictly less than $C$, there is a unique index $i$ such that $c_i$ is absent from $S$, while $c_k, ..., c_{i+1}$ are present in $S$, and no other value larger than $c_i$ is. One can therefore group those $k$-combinations $S$ according to the possible values 1, 2, ..., $k$ of $i$, and count each group separately. For a given value of $i$ one must include $c_k, ..., c_{i+1}$ in $S$, and the remaining $i$ elements of $S$ must be chosen from the $c_i$ non-negative integers strictly less than $c_i$; moreover any such choice will result in a $k$-combinations $S$ strictly less than $C$. The number of possible choices is $\binom{c_i}{i}$, which is therefore the number of combinations in group $i$; the total number of $k$-combinations strictly less than $C$ then is,

$$\binom{c_1}{1} + \binom{c_2}{2} + \cdots + \binom{c_k}{k},$$

and this is the index (starting from 0) of $C$ in the ordered list of $k$-combinations. Obviously there is for every $N \in$ N exactly one $k$-combination at index $N$ in the list (supposing $k \geq 1$, since the list is then infinite), so the above argument proves that every $N$ can be written in exactly one way as a sum of $k$ binomial coefficients of the given form.

## Finding the k-combination for a Given Number

The given formula allows finding the place in the lexicographic ordering of a given $k$-combination immediately. The reverse process of finding the $k$-combination at a given place $N$ requires somewhat more work, but is straightforward nonetheless. By the definition of the lexicographic ordering, two $k$-combinations that differ in their largest element $c_k$ will be ordered according to the comparison of those largest elements, from which it follows that all combinations with a fixed value of their largest element are contiguous in the list. Moreover the smallest combination with $c_k$ as largest element is $\binom{c_k}{k}$, and it has $c_i = i - 1$ for all $i < k$ (for this combination all terms in the expression except $\binom{c_k}{k}$ are zero). Therefore $c_k$ is the largest number such that $\binom{c_k}{k} \leq N$. If $k > 1$ the remaining elements of the $k$-combination form the $k - 1$-combination corresponding to the number $N - \binom{c_k}{k}$ in the combinatorial number system of degree $k - 1$, and can therefore be found by continuing in the same way for $N - \binom{c_k}{k}$ and $k - 1$ instead of $N$ and $k$.

Example:

Suppose one wants to determine the 5-combination at position 72. The successive values of $\binom{n}{5}$ for $n = 4, 5, 6, \ldots$ are 0, 1, 6, 21, 56, 126, 252, ..., of which the largest one not exceeding 72 is 56, for $n = 8$. Therefore $c_5 = 8$, and the remaining elements form the 4-combination at position $72 - 56 = 16$. The successive values of $\binom{n}{4}$ for $n = 3, 4, 5, \ldots$ are 0, 1, 5, 15, 35, ..., of which the largest one not exceeding 16 is 15, for $n = 6$, so $c_4 = 6$. Continuing similarly to search for a 3-combination at position $16 - 15 = 1$ one finds $c_3 = 3$, which uses up the final unit; this establishes $72 = \binom{8}{5} + \binom{6}{4} + \binom{3}{3}$, and the remaining values $c_i$ will be the maximal ones with $\binom{c_i}{i} = 0$, namely $c_i = i - 1$. Thus we have found the 5-combination {8, 6, 3, 1, 0}.

## National Lottery Example in Excel

For each of the $\binom{49}{6}$ lottery combinations $c_1 < c_2 < c_3 < c_4 < c_5 < c_6$, there is a list number $N$ between 0 and $\binom{49}{6} - 1$ which can be found by adding:

$$\binom{49 - c_1}{6} + \binom{49 - c_2}{5} + \binom{49 - c_3}{4} + \binom{49 - c_4}{3} + \binom{49 - c_5}{2} + \binom{49 - c_6}{1}.$$

Suppose you want to find a position of a British National Lottery result 3,6,15,17,18,35 in a list of possible results. Excel function COMBIN(49,6) would show that number of results is 13983816. Now if you put numbers 3,6,15,17,18,35 each in its cell and formulas COMBIN(49-3,6), COMBIN(49-6,5), COMBIN(49-15,4), COMBIN(49-17,3), COMBIN(49-18,2), 49−35 under each of them. Use cell references instead of actual values, the actual values are provided for readability. You would get a row of numbers of 9366819,962598,46376,4960,465,14. Adding these would show particular position 10381232. Note that you do not need use formula COMBIN(49-35,1) to get 14. You can have it by subtraction 49-35 alone. Also the COMBIN function does not return 0 in case 49-X becomes less than 6. You'd need to use IF with ISNUMBER function to convert #NUM! into 0.

Now the reverse engineering is a bit trickier. You could use 49 IF statements in one cell or use solver to find maximum argument for COMBIN result to be less or equal than position number.

Instead let's use a table of 6 by 49 and use MATCH function where resulting row number would be the argument and a ball number. If you make column headings from 6 to 1 (B1:G1) in descending order and row labels of 1 to 49 (A2:A50) in ascending order (vertically ascending in Excel means numbers growing from top to bottom). Then if you fill up the table with formula COMBIN($A2,B$1) starting from left top corner. $ signs will make sure that index values are always taken from heading row and label column. Replace #NUM! with zeros. You should get something like this:

| | 6 | 5 | 4 | 3 | 2 | 1 |
|---|---|---|---|---|---|---|
| 1 | 0 | 0 | 0 | 0 | 0 | 1 |
| 2 | 0 | 0 | 0 | 0 | 1 | 2 |
| 3 | 0 | 0 | 0 | 1 | 3 | 3 |
| 4 | 0 | 0 | 1 | 4 | 6 | 4 |
| 5 | 0 | 1 | 5 | 10 | 10 | 5 |
| 6 | 1 | 6 | 15 | 20 | 15 | 6 |
| 7 | 7 | 21 | 35 | 35 | 21 | 7 |
| 8 | 28 | 56 | 70 | 56 | 28 | 8 |
| 9 | 84 | 126 | 126 | 84 | 36 | 9 |
| 10 | 210 | 252 | 210 | 120 | 45 | 10 |
| 11 | 462 | 462 | 330 | 165 | 55 | 11 |
| 12 | 924 | 792 | 495 | 220 | 66 | 12 |
| ... | | | | | | |
| 49 | 13983816 | 1906884 | 211876 | 18424 | 1176 | 49 |

Now if you create a new row of six cells and fill them with formulas which would find the largest values in each column which are less than or equal to position number in question: The first cell with =INDEX(B2:B50,MATCH(10381232,B2:B50)), the rest of the cells with

```
INDEX(C2:C50,MATCH(10381232-SUM(of previous cells),C2:C50))

...

INDEX(G2:G50,MATCH(10381232-SUM(of previous cells),G2:G50))
```

This would present you with a row of values you have already seen 9366819,962598,46376,4960, 465,14, Now in a next row first cell write =49-MATCH(10381232,B2:B50) and similarly,

```
=49-MATCH(10381232-9366819,C2:C50)

...

=49-MATCH(10381232-9366819-962598-46376-4960-465,G2:G50)
```

Again use the references to cells instead of actual values. You should be presented with original ball numbers of 3,6,15,17,18,35.

Now you can generate a fresh Lottery number combination from single = randbetween (1,combin (49,6)) or you could look at the list position numbers of earlier results to see if there is a trend.

## Applications

One could use the combinatorial number system to list or traverse all $k$-combinations of a given finite set, but this is a very inefficient way to do that. Indeed, given some $k$-combination it is much easier to find the next combination in lexicographic ordering directly than to convert a number to a $k$-combination. To find the next combination, find the smallest $i \geq 2$ for which $c_i \geq c_{i-1}+2$ (if there is no such $i$, take $i = k+1$); then increase $c_{i-1}$ by one and set all $c_j$ with $j < i - 1$ to their minimal value $j - 1$. If the $k$-combination is represented by its characteristic vector, that is as a binary value with $k$ bits 1, then the next such value can be computed without any loop using bitwise arithmetic: the following C++ function will advance x to that value or return false:

```
// find next k-combination

bool next_combination(unsigned long& x) // assume x has form x'01^a10^b in binary

{

  unsigned long u = x & -x; // extract rightmost bit 1; u = 0'00^a10^b

  unsigned long v = u + x; // set last non-trailing bit 0, and clear to the right;
v=x'10^a00^b

  if (v==0) // then overflow in v, or x==0

    return false; // signal that next k-combination cannot be represented

   x = v +(((v^x)/u)>>2); // v^x = 0'11^a10^b, (v^x)/u = 0'0^b1^{a+2}, and x ←
x'100^b1^a

  return true; // successful completion

}
```

This is called Gosper's hack; corresponding assembly code was described as item 175 in HAKMEM.

On the other hand the possibility to directly generate the $k$-combination at index $N$ has useful applications. Notably, it allows generating a random $k$-combination of an $n$-element set using a random integer $N$ with $0 \leq N < \binom{n}{k}$, simply by converting that number to the corresponding $k$-combination. If a computer program needs to maintain a table with information about every $k$-combination of a given finite set, the computation of the index $N$ associated with a combination will allow the table to be accessed without searching.

# Combinatorial Modelling

Combinatorial modelling is the process which lets us identify a suitable mathematical model to reformulate a problem. These combinatorial models will provide, through the combinatorics theory, the operations needed to solve the problem.

## Implicit Combinatorial Models

Simple combinatorial problems are the ones that can be solved by applying just one combinatorial operation (variations, permutations, combinations, etc.). These problems can be classified into three different models, called implicit combinatorial models.

## Selection

A selection problem requires to choose a sample of $k$ elements out of a set of $n$ elements. It is needed to know if the order in which the objects are selected matters and whether an object can be selected more than once or not. This table shows the operations that the model provides to get the number of different samples for each of the selections:

| | Repetition not allowed | Repetition allowed |
|---|---|---|
| Ordered sample | $V_{n,k}$ | $VR_{n,k}$ |
| Non ordered sample | $C_{n,k}$ | $CR_{n,k}$ |

## Examples

1. At a party there are 50 people. Everybody shakes everybody's hand once. How often are *hands shaken in total?* What we need to do is calculate the number of all possible pairs of party guests, which means, a sample of 2 people out of the 50 guests, so $k=2$ and $n=50$. A pair will be the same no matter the order of the two people. A handshake must be carried out by two different people (no repetition). So, it is required to select an ordered sample of 2 elements out of a set of 50 elements, in which repetition is not allowed. That is all we need to know to choose the right operation, and the result is:

$$C_{50,2} = \frac{50!}{2!\cdot(50-2)!} = 1225$$

2. Unfortunately, you can't remember the code for your four-digit lock. You only know that you didn't use any digit more than once. How many different ways do you have to try? We need to choose a sample of 4 digits out of the set of 10 digits (base 10), so $k=4$ and $n=10$. The digits must be ordered in a certain way to get the correct number, so we want to select an ordered sample. As the statement says, no digit was chosen more than once, so our sample will not have repeated digits. So, it is required to select an ordered sample of 4 elements out of a set of 10 elements, in which repetition is not allowed. That is all we need to know to choose the right operation, and the result is:

$$V_{10,4} = \frac{10!}{(10-4)!} = 5040$$

3. A boy wants to buy 20 invitation cards to give to their friends for his birthday party. There are 3 types of cards in the store, and he likes them all. In how many ways can he buy the 20 cards? It is required to choose a sample of 20 invitation cards out of the set of 3 types of

cards, so $k = 20$ and $n = 3$. The order in which you choose the different types of invitations does not matter. As a type of card must be selected more than once, there will be repetitions in our invitation cards. So, we want to select a non ordered sample of 20 elements ($n = 3$ ) out of a set of 3 elements ($k = 20$), in which repetition is allowed. That is all we need to know to choose the right operation, and the result is:

$$CR_{3,20} = C_{22,20} = \frac{22!}{20! \cdot (22-20)!} = 231$$

## Distribution

In a distribution problem it is required to place *k* objects into *n* boxes or recipients. In order to choose the right operation out of the ones that the model provides, it is necessary to know:

- Whether the objects are distinguishable or not.
- Whether the boxes are distinguishable or not.
- If the order in which the objects are placed in a box matters.
- The conditions that the distribution must meet. Depending on this, the distribution can be:
  - Injective distribution: every box must have at most 1 object ($k \leq n$).
  - Surjective distribution: every box must have at least 1 object ($k \geq n$).
  - Bijective distribution: every box must exactly 1 object ($k = n$).
  - Distribution with no restrictions.

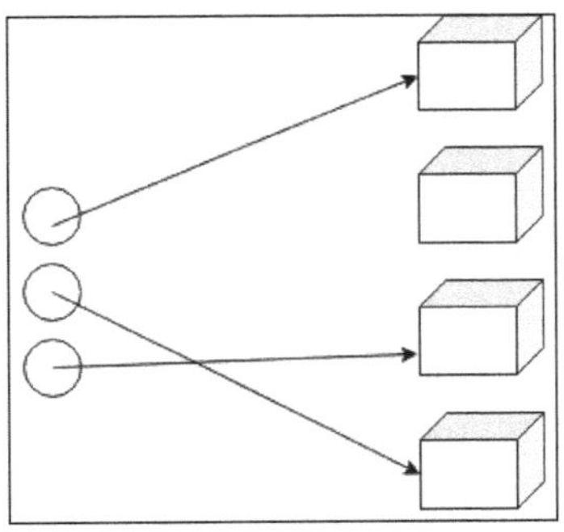

Injective distribution.

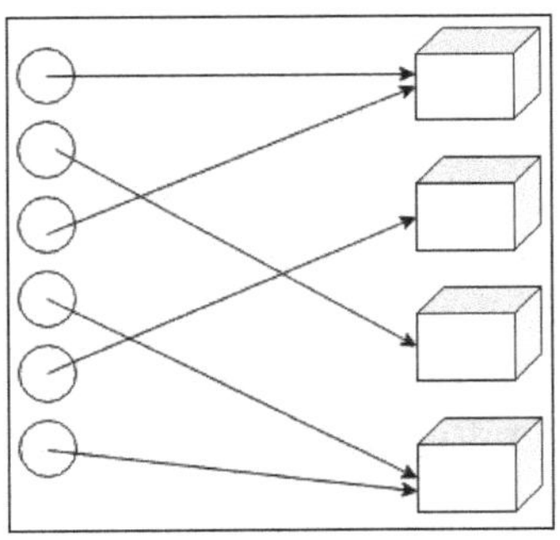

Surective distribution.

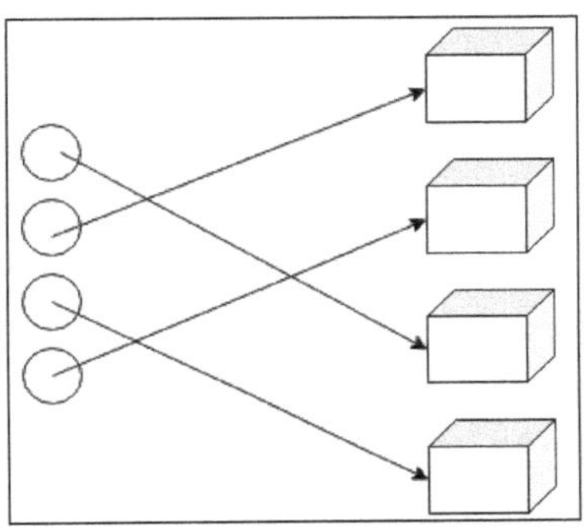

Bijective distribution.

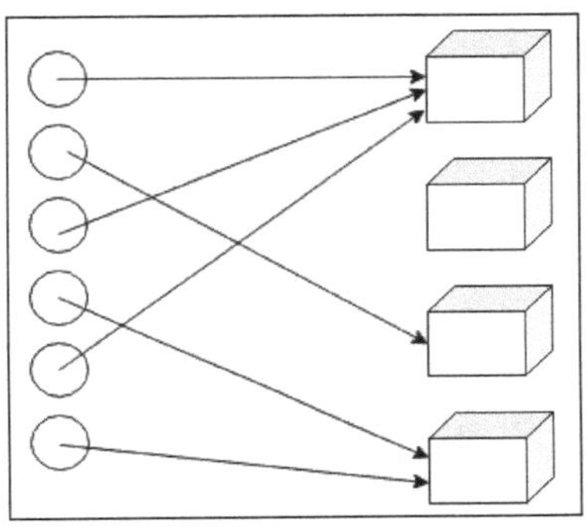

Distribution with no restrictions.

The following table shows the operations that the model provides to get the number of different ways of distributing the objects for each of the distributions:

| Distribution of k objects into n boxes | | | | | | |
|---|---|---|---|---|---|---|
| | Non ordered distribution | | | | Ordered distribution | |
| | Distinguishable objects | | Indistinguishable objects | | Distinguishable objects | |
| | Distinguish-able boxes | Indistinguish-able boxes | Distinguish-able boxes | Indistinguish-able boxes | Distinguish-able boxes | Indistinguishable boxes |
| Any distribution | $VR_{n,k}$ | $\sum_{i=1}^{n} S(k,i)$ | $CR_{n,k}$ | $\sum_{i=1}^{n} P(k,i)$ | $k!\cdot CR_{n,k}$ | $\sum_{i=1}^{n} L(k,i)$ |
| Injective distribution | $V_{n,k}$ | 1 | $C_{n,k}$ | 1 | $V_{n,k}$ | 1 |
| Surjective distribution | $n!\cdot S(k,n)$ | $S(k,n)$ | $CR_{n,k-n}$ | $P(k,n)$ | $k!\cdot CR_{n,k-n}$ | $L(k,n)=\frac{k!}{n!}\cdot\binom{k-1}{n-1}$ |
| Bijective distribution | $n!$ | 1 | 1 | 1 | $n!$ | 1 |

$S(k,n)=\left\{ \begin{matrix} k \\ n \end{matrix} \right\}$: Stirling numbers of the second kind.

$P(k,n)$: Number of partitions of the integer $k$ into $n$ parts.

$L(k,n)$: Lah numbers (Stirling numbers of the third kind).

## Examples

1. A maths teacher has to give 3 studentships among his students. 7 of them got an 'outstanding' grade, so they are the candidates to get them. In how many ways can he distribute the grants? Let's consider the 3 studentships are objects that have to be distributed into 7 boxes, which are the students. As the objects are identical studentships, they are indistinguishable. The boxes are distinguishable, as they are different students. Every studentship must be given to a different student, so every box must have at most 1 object. Furthermore, the order in which the objects are placed in a boxes does not matter, because there cannot be more than one on each box. So, it is a non ordered injective distribution of 3 indistinguishable objects ($k=3$) into 7 distinguishable boxes ($n=7$). That is all we need to know to choose the right operation, and the result is:

$$C_{7,3}=\frac{7!}{3!\cdot(7-3)!}=35$$

2. A group of 8 friends rent a 5-room cottage to spend their holidays. If the rooms are identical and no one can be empty, in how many ways can they be distributed in the cottage? Let's consider the friends are objects that have to be distributed into 5 boxes, which are the rooms. As the objects are different people, they are distinguishable. The boxes are indistinguishable, as they are identical rooms. We can consider it as a non ordered distribution,

because the ordered in which everyone is placed in the rooms does not matter. No room can be empty, so every box must have at least 1 object. So, it is a non ordered surjective distribution of 8 distinguishable objects ($k = 8$) into 5 indistinguishable boxes ($n = 5$). That is all we need to know to choose the right operation, and the result is:

$$S(8,5) = \left\{ \begin{matrix} 8 \\ 5 \end{matrix} \right\} = \frac{1}{5!} \sum_{i=0}^{5} (-1)^i \binom{5}{i} (5-i)^8 = 1050$$

3. 12 people are done shopping in a supermarket where 4 cashiers are working at the moment. In how many different ways can they be distributed into the checkouts? Let's consider the people are objects that have to be distributed into boxes, which are the check-outs. As the people and the checkouts are different, the objects and the boxes are distinguishable. The order in which the objects are placed in the boxes matter, because they are people getting into queues. The statement does not mention any restriction. So, it is an ordered distribution with no restrictions of 12 distinguishable objects ($k = 12$) into 4 distinguishable boxes ($n = 4$). That is all we need to know to choose the right operation, and the result is:

$$12! \cdot CR_{4,12} = 12! \cdot C_{15,12} = 12! \cdot \frac{15!}{12! \cdot (15-12)!} = 217945728000$$

## Partition

A partition problem requires to divide a set of *k* elements into *n* subsets. This model is related to the distribution one, as we can consider the objects inside every box as subsets of the set of objects to distribute. So, each of the 24 distributions described previously has a matching kind of partition into subsets. So, a partition problem can be solved by transforming it into a distribution one and applying the correspondent operation provided by the distribution model. Following this method, we will get the number of possible ways of dividing the set. The relation between these two models is described in the following table:

| Distribution | Partition |
|---|---|
| Ordered | Subsets of ordered elements |
| Non ordered | Subsets of non ordered elements |
| Distinguishable objects | Distinguishable elements |
| Indistinguishable objects | Indistinguishable elements |
| Distinguishable boxes | Ordered partitions |
| Indistinguishable boxes | Non ordered partitions |
| Injective distribution | Subsets of 1 element or empty ones |
| Surjective distribution | No empty subsets |
| Bijective distribution | Subsets of exactly 1 element |
| No restrictions | Subsets of 1 or more elements or empty ones |

This relation let us transform the table provided by the distribution model into a new one that can be used to know the different ways of dividing a set of $k$ elements into $n$ subsets:

| Partition of a set of k elements into n subsets | | | | | | |
|---|---|---|---|---|---|---|
| | Non ordered elements | | | | , | |
| | Distinguishable elements | | Indistinguishable elements | | Distinguishable elements | |
| | Ordered subsets | Non ordered subsets | Ordered subsets | Non ordered subsets | Ordered subsets | Non ordered subsets |
| Any subsets | $VR_{n,k}$ | $\sum_{i=1}^{n} S(k,i)$ | $CR_{n,k}$ | $\sum_{i=1}^{n} P(k,i)$ | $k!\cdot CR_{n,k}$ | $\sum_{i=1}^{n} L(k,i)$ |
| Empty or 1-element subsets | $V_{n,k}$ | 1 | $C_{n,k}$ | 1 | $V_{n,k}$ | 1 |
| No empty subsets | $n!\cdot S(k,n)$ | $S(k,n)$ | $CR_{n,k-n}$ | $P(k,n)$ | $k!\cdot CR_{n,k-n}$ | $L(k,n)=\frac{k!}{n!}\cdot CR_{n,k-n}$ |
| Subsets of 1 element | $n!$ | 1 | 1 | 1 | $n!$ | 1 |

Examples: A group of 3 classmates have to make a thesis about 8 different maths topics. In how many ways can they split the work to do? We need to divide the set of 8 topics into 3 subsets. These subsets will be the topics that each of the students will work on. The elements in the set (topics) are distinguishable. The partitions must be ordered because each one will correspond to a different student, but the topics inside every subset do not have to be ordered because each student can decide which order to follow when working on the thesis. The statement does not mention any restriction of the subsets. So, it is required to divide a set of 8 elements ($k=8$) into 3 ordered subsets ($n=3$) of non ordered elements. That is all we need to know to choose the right operation, and the result is:

$$VR_{3,8} = 3^8 = 6561$$

## Pascal's Identity

Pascal's Identity is a useful theorem of combinatorics dealing with combinations (also known as binomial coefficients). It can often be used to simplify complicated expressions involving binomial coefficients.

Pascal's Identity is also known as Pascal's Rule, Pascal's Formula, and occasionally Pascal's Theorem.

Theorem: Pascal's Identity states that,

$$\binom{n}{k} = \binom{n-1}{k-1} + \binom{n-1}{k}$$

for any positive integers $k$ and $n$. Here, $\binom{n}{k}$ is the binomial coefficient $\binom{n}{k} = nCK = C_k^n$.

This result can be interpreted combinatorially as follows: the number of ways to choose $k$ things from $n$ things is equal to the number of ways to choose $k-1$ things from things added to the number of ways to choose $k$ things from $n-1$ things.

Proof: If $k > n$ then $\binom{n}{k} = 0 = \binom{n-1}{k-n} + \binom{n-1}{k}$ and so the result is trivial. So assume $k \le n$. Then:

$$\begin{aligned}\binom{n-1}{k-1} + \binom{n-1}{k} &= \frac{(n-1)!}{(k-1)!(n-k)!} + \frac{(n-1)!}{k!(n-k-1)!} \\ &= (n-1)!\left(\frac{k}{k!(n-k)!} + \frac{n-k}{k!(n-k)!}\right) \\ &= (n-1)!.\frac{n}{k!(n-k)!} \\ &= \frac{n!}{k!(n-k)!} \\ &= \binom{n}{k}.\end{aligned}$$

Alternate Proofs: Here, we prove this using committee forming.

Consider picking one fixed object out of $n$ objects. Then, we can choose $k$ objects including that one in $\binom{n-1}{k-1}$ ways.

Because our final group of objects either contains the specified one or doesn't, we can choose the group in $\binom{n-1}{k-1} + \binom{n-1}{k}$ ways.

But we already know they can be picked in $\binom{n}{k}$ ways, so:

$$\binom{n}{k} = \binom{n-1}{k-1} + \binom{n-1}{k}$$

Also, we can look at Pascal's Triangle to see why this is. If we were to extend Pascal's Triangle to row n, we would see the term $\binom{n}{k}$. Above that, we would see the terms $\binom{n-1}{k-1}$ and $\binom{n-1}{k}$. Due to the definition of Pascal's Triangle, $\binom{n}{k} = \binom{n-1}{k-1} + \binom{n-1}{k}$.

# Permutation

In mathematics, permutation is the act of arranging the members of a set into a sequence or order, or, if the set is already ordered, rearranging (reordering) its elements—a process called permuting. Permutations differ from combinations, which are selections of some members of a set regardless of order. For example, written as tuples, there are six permutations of the set {1,2,3}, namely: (1,2,3), (1,3,2), (2,1,3), (2,3,1), (3,1,2), and (3,2,1). These are all the possible orderings of this three-element set. Anagrams of words whose letters are different are also permutations: the letters are already ordered in the original word, and the anagram is a reordering of the letters. The study of permutations of finite sets is an important topic in the fields of combinatorics and group theory.

Each of the six rows is a different permutation of three distinct balls.

Permutations are studied in almost every branch of mathematics, and in many other fields of science. In computer science, they are used for analyzing sorting algorithms; in quantum physics, for describing states of particles; and in biology, for describing RNA sequences.

The number of permutations of $n$ distinct objects is $n$ factorial, usually written as $n!$, which means the product of all positive integers less than or equal to $n$.

In algebra, and particularly in group theory, a permutation of a set $S$ is defined as a bijection from $S$ to itself. That is, it is a function from $S$ to $S$ for which every element occurs exactly once as an image value. This is related to the rearrangement of the elements of $S$ in which each element $s$ is replaced by the corresponding $f(s)$. For example, the permutation (3,1,2) mentioned above is described by the function $\alpha$ defined as:

$$\alpha(1) = 3, \quad \alpha(2) = 1, \quad \alpha(3) = 2.$$

The collection of such permutations form a group called the symmetric group of $S$. The key to this group's structure is the fact that the composition of two permutations (performing two given rearrangements in succession) results in another rearrangement. Permutations may act on structured objects by rearranging their components, or by certain replacements (substitutions) of symbols. Frequently the set used is $S = \mathbb{N}_n = \{1, 2, \ldots, n\}$,, but there is no restriction on the set.

In elementary combinatorics, the $k$-permutations, or partial permutations, are the ordered arrangements of $k$ distinct elements selected from a set. When $k$ is equal to the size of the set, these are the permutations of the set.

In mathematics texts it is customary to denote permutations using lowercase greek letters. Commonly, either $\alpha$ and $\beta$, or $\sigma, \tau$ and $\pi$ are used.

Permutations can be defined as bijections from a set $S$ onto itself. All permutations of a set with $n$ elements form a symmetric group, denoted $S_n$, where the group operation is function composition. Thus for two permutations, $\pi$ and $\sigma$ in the group $S_n$, the four group axioms hold:

1. Closure: If $\pi$ and $\sigma$ are in $S_n$ then so is $\pi\sigma$.
2. Associativity: For any three permutations $\pi, \sigma, \tau \in S_n$, $(\pi\sigma)\tau = \pi(\sigma\tau)$.
3. Identity: There is an identity permutation, denoted id and defined by $\mathrm{id}(x) = x$ for all $x \in S$. For any $\sigma \in S_n$ , $\mathrm{id}\,\sigma = \sigma\,\mathrm{id} = \sigma$.
4. Invertibility: For every permutation $\pi \in S_n$, there exists $\pi^{-1} \in S_n$ so that $\pi\pi^{-1} = \pi^{-1}\pi = \mathrm{id}$.

In general composition of two permutations is not commutative, $\pi\sigma \neq \sigma\pi$.

In this point of view it is emphasized that a permutation is a function that *performs* a rearrangement of a set and is not a rearrangement itself. The older and more elementary view holds that permutations are the rearrangements. Sometimes to distinguish these two views, the words *active* and *passive* forms, or in older terminology *substitutions* and *permutations* are used.

A permutation can be decomposed into one or more disjoint *cycles*, that is, the orbits, which are found by repeatedly tracing the application of the permutation on some elements. For example, the permutation $\sigma$ defined by $\sigma(7) = 7$ has a 1-cycle, $(7)$ while the permutation $\pi$ defined by $\pi(2) = 3$ and $\pi(3) = 2$ has a 2-cycle $(23)$. In general, a cycle of length $k$, that is, consisting of $k$ elements, is called a $k$-cycle.

An element in a 1-cycle $(x)$ is called a fixed point of the permutation. A permutation with no fixed points is called a derangement. 2-cycles are called transpositions; such permutations merely exchange the place of two elements, leaving the others fixed.

## Notations

Since writing permutations elementwise, that is, as piecewise functions, is cumbersome, several notations have been invented to represent them more compactly. *Cycle notation* is a popular choice for many mathematicians due to its compactness and the fact that it makes a permutation's structure transparent. It is the notation used in this article unless otherwise specified, but other notations are still widely used, especially in application areas.

### Two-line Notation

In Cauchy's *two-line notation*, one lists the elements of $S$ in the first row, and for each one its image below it in the second row. For instance, a particular permutation of the set $S = \{1,2,3,4,5\}$ can be written as:

$$\sigma = \begin{pmatrix} 1 & 2 & 3 & 4 & 5 \\ 2 & 5 & 4 & 3 & 1 \end{pmatrix};$$

this means that $\sigma$ satisfies $\sigma(1) = 2$, $\sigma(2) = 5$, $\sigma(3) = 4$, $\sigma(4) = 3$, and $\sigma(5) = 1$. The elements of $S$ may appear in any order in the first row. This permutation could also be written as:

$$\sigma = \begin{pmatrix} 3 & 2 & 5 & 1 & 4 \\ 4 & 5 & 1 & 2 & 3 \end{pmatrix}.$$

## One-line Notation

If there is a "natural" order for the elements of $S$, say $x_1, x_2, \ldots, x_n$, then one uses this for the first row of the two-line notation:

$$\sigma = \begin{pmatrix} x_1 & x_2 & x_3 & \cdots & x_n \\ \sigma(x_1) & \sigma(x_2) & \sigma(x_3) & \cdots & \sigma(x_n) \end{pmatrix}.$$

Under this assumption, one may omit the first row and write the permutation in *one-line notation* as,

$$(\sigma(x_1)\,\sigma(x_2)\,\sigma(x_3)\cdots\sigma(x_n)),$$

that is, an ordered arrangement of S. Care must be taken to distinguish one-line notation from the cycle notation In mathematics literature, a common usage is to omit parentheses for one-line notation, while using them for cycle notation. The one-line notation is also called the *word representation* of a permutation. The example above would then be 2 5 4 3 1 since the natural order 1 2 3 4 5 would be assumed for the first row. (It is typical to use commas to separate these entries only if some have two or more digits.) This form is more compact, and is common in elementary combinatorics and computer science. It is especially useful in applications where the elements of $S$ or the permutations are to be compared as larger or smaller.

## Cycle Notation

Cycle notation describes the effect of repeatedly applying the permutation on the elements of the set. It expresses the permutation as a product of cycles; since distinct cycles are disjoint, this is referred to as "decomposition into disjoint cycles".

To write down the permutation $\sigma$ in cycle notation, one proceeds as follows:

1. Write an opening bracket then select an arbitrary element $x$ of $S$ and write it down: $(x$
2. Then trace successive applications of $\sigma$ write it down: $(x\sigma(x)\sigma(\sigma(x))\ldots$
3. Repeat until the value returns to $x$ and write down a closing parenthesis rather than $x$: $(x\sigma(x)\sigma(\sigma(x))\ldots)$

4. Now continue with an element $y$ of $S$, not yet written down, and proceed in the same way: $(x\sigma(x)\sigma(\sigma(x))\ldots)(y\ldots)$
5. Repeat until all elements of $S$ are written in cycles.

Since for every new cycle the starting point can be chosen in different ways, there are in general many different cycle notations for the same permutation; for the example above one has:

$$\begin{pmatrix} 1 & 2 & 3 & 4 & 5 \\ 2 & 5 & 4 & 3 & 1 \end{pmatrix} = (1\ \ 2\ \ 5)(3\ \ 4) = (3\ \ 4)(1\ \ 2\ \ 5) = (3\ \ 4)(5\ \ 1\ \ 2).$$

1-cycles are often omitted from the cycle notation, provided that the context is clear; for any element $x$ in $S$ not appearing in any cycle, one implicitly assumes $\sigma(x) = x$. The identity permutation, which consists only of 1-cycles, can be denoted by a single 1-cycle (x), by the number 1, or by *id*.

A convenient feature of cycle notation is that one can find a permutation's inverse simply by reversing the order of the elements in the permutation's cycles. For example:

$$[(125)(34)]^{-1} = (521)(43)$$

## Canonical Cycle Notation

In some combinatorial contexts it useful to fix a certain order for the elements in the cycles and of the (disjoint) cycles themselves. Miklós Bóna calls the following ordering choices the *canonical cycle notation*:

- In each cycle the *largest* element is listed first.
- The cycles are sorted in *increasing* order of their first element.

For example, (312)(54)(8)(976) is a permutation in canonical cycle notation. The canonical cycle notation does not omit one-cycles.

Richard P. Stanley calls the same choice of representation the "standard representation" of a permutation. and Martin Aigner uses the term "standard form" for the same notion. Sergey Kitaev also uses the "standard form" terminology, but reverses both choices; i.e., each cycle lists its least element first and the cycles are sorted in decreasing order of their least, that is, first elements.

## Composition of Permutations

There are two ways to denote the composition of two permutations. $\sigma\cdot\pi$ is the function that maps any element $x$ of the set to $\sigma(\pi(x))$. The rightmost permutation is applied to the argument first, because of the way function application is written.

Since function composition is associative, so is the composition operation on permutations: $(\sigma\pi)\tau = \sigma(\pi\tau)$. Therefore, products of more than two permutations are usually written without adding parentheses to express grouping; they are also usually written without a dot or other sign to indicate composition.

## Other Uses of the Term Permutation

The concept of a permutation as an ordered arrangement admits several generalizations that are not permutations but have been called permutations in the literature.

## *k*-permutations of *n*

A weaker meaning of the term *permutation*, sometimes used in elementary combinatorics texts, designates those ordered arrangements in which no element occurs more than once, but without the requirement of using all the elements from a given set. These are not permutations except in special cases, but are natural generalizations of the ordered arrangement concept. Indeed, this use often involves considering arrangements of a fixed length $k$ of elements taken from a given set of size $n$, in other words, these k-permutations of n are the different ordered arrangements of a $k$-element subset of an $n$-set (sometimes called variations in the older literature.) These objects are also known as partial permutations or as sequences without repetition, terms that avoid confusion with the other, more common, meaning of "permutation". The number of such $k$-permutations of $n$ is denoted variously by such symbols as $P_k^n$, ${}_nP_k$, ${}^nP_k$, $P_{n,k}$, or $P(n,k)$, and its value is given by the product,

$$P(n,k) = \underbrace{n \cdot (n-1) \cdot (n-2) \cdots (n-k+1)}_{k \text{ factors}},$$

which is 0 when $k > n$, and otherwise is equal to:

$$\frac{n!}{(n-k)!}.$$

The product is well defined without the assumption that $n$ is a non-negative integer and is of importance outside combinatorics as well; it is known as the Pochhammer symbol $(n)_k$ or as the $k$-th falling factorial power $n^{\underline{k}}$ of $n$.

This usage of the term *permutation* is closely related to the term *combination*. A $k$-element combination of an $n$-set $S$ is a $k$ element subset of $S$, the elements of which are not ordered. By taking all the $k$ element subsets of $S$ and ordering each of them in all possible ways we obtain all the $k$-permutations of $S$. The number of $k$-combinations of an $n$-set, $C(n,k)$, is therefore related to the number of $k$-permutations of $n$ by:

$$C(n,k) = \frac{P(n,k)}{P(k,k)} = \frac{\frac{n!}{(n-k)!}}{\frac{k!}{0!}} = \frac{n!}{(n-k)!k!}.$$

These numbers are also known as binomial coefficients and denoted $\binom{n}{k}$.

## Permutations with Repetition

Ordered arrangements of the elements of a set $S$ of length $n$ where repetition is allowed are called $n$-tuples, but have sometimes been referred to as permutations with repetition although they are not permutations in general. They are also called words over the alphabet $S$ in some contexts. If the set $S$ has $k$ elements, the number of $n$-tuples over $S$ is: $k^n$.

There is no restriction on how often an element can appear in an $n$-tuple, but if restrictions are placed on how often an element can appear, this formula is no longer valid.

## Permutations of Multisets

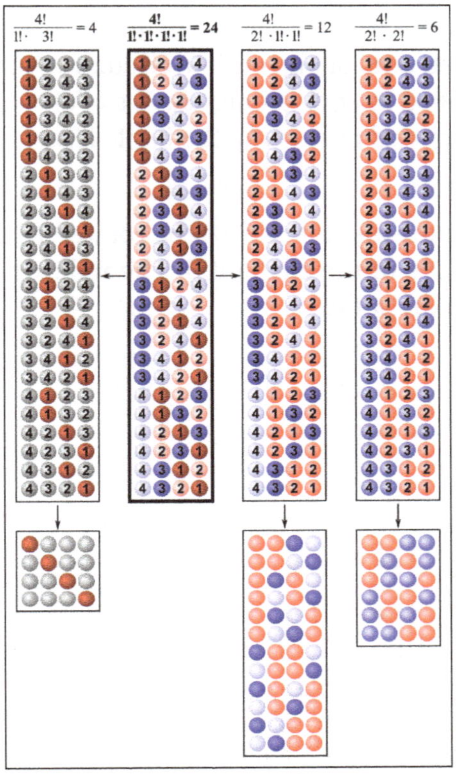


Permutations of multisets.

If $M$ is a finite multiset, then a multiset permutation is an ordered arrangement of elements of $M$ in which each element appears exactly as often as is its multiplicity in $M$. An anagram of a word having some repeated letters is an example of a multiset permutation. If the multiplicities of the elements of $M$ (taken in some order) are $m_1$, $m_2$, ..., $m_l$ and their sum (i.e., the size of $M$) is $n$, then the number of multiset permutations of $M$ is given by the multinomial coefficient,

$$\binom{n}{m_1, m_2, \ldots, m_l} = \frac{n!}{m_1!\,m_2!\cdots m_l!} = \frac{\left(\sum_{i=1}^{l} m_i\right)!}{\prod_{i=1}^{l} m_i!}.$$

For example, the number of distinct anagrams of the word MISSISSIPPI is:

$$\frac{11!}{1!4!4!2!} = 34650$$

A k-permutation of a multiset $M$ is a sequence of length $k$ of elements of $M$ in which each element appears *at most* its multiplicity in M times (an element's repetition number).

## Circular Permutations

Permutations, when considered as arrangements, are sometimes referred to as *linearly ordered* arrangements. In these arrangements there is a first element, a second element, and so on. If, however, the objects are arranged in a circular manner this distinguished ordering no longer exists, that is, there is no "first element" in the arrangement, any element can be considered as the start of the arrangement. The arrangements of objects in a circular manner are called circular permutations. These can be formally defined as equivalence classes of ordinary permutations of the objects, for the equivalence relation generated by moving the final element of the linear arrangement to its front.

Two circular permutations are equivalent if one can be rotated into the other (that is, cycled without changing the relative positions of the elements). The following two circular permutations on four letters are considered to be the same.

```
     1               4
4        3       2       1
     2               3
```

The circular arrangements are to be read counterclockwise, so the following two are not equivalent since no rotation can bring one to the other.

```
     1               1
4        3       3       4
     2               2
```

The number of circular permutations of a set $S$ with $n$ elements is $(n-1)!$.

## Properties

The number of permutations of $n$ distinct objects is $n!$.

The number of $n$-permutations with $k$ disjoint cycles is the signless Stirling number of the first kind, denoted by $c(n, k)$.

## Permutation Type

The cycles of a permutation partition the set $S_n$ so the lengths of the cycles of a permutation $\sigma$ forms a partition of $n$ called the cycle type of $\sigma$. There is a "1" in the cycle type for every fixed point of $\sigma$, a "2" for every transposition, and so on. The cycle type of $\beta = (125)(34)(68)(7)$ is $(3,2,2,1)$ which is sometimes written in a more compact form as $[1^1 2^2 3^1]$.

The general form is $[1^{\alpha_1} 2^{\alpha_2} \ldots n^{\alpha_n}]$, where $\alpha_1, \ldots, \alpha_n$ are the number of cycles of respective length. The number of partitions of a certain type is then given by:

$$\frac{n!}{1^{\alpha_1} 2^{\alpha_2} \ldots n^{\alpha_n} \alpha_1! \alpha_2! \ldots \alpha_n!}.$$

## Conjugating Permutations

In general, composing permutations written in cycle notation follows no easily described pattern – the cycles of the composition can be different from those being composed. However the cycle structure is preserved in the special case of conjugating a permutation $\sigma$ by another permutation $\pi$, which means forming the product $\pi\sigma\pi^{-1}$. Here, $\pi\sigma\pi^{-1}$ is the *conjugate* of $\sigma$ and its cycle notation can be obtained by taking the cycle notation for $\sigma$ and applying $\pi$ to all the entries in it. It follows that two permutations are conjugate exactly when they have the same type.

## Permutation Order

The order of a permutation $\sigma$ is the smallest positive integer $m$ so that $\sigma^m = \mathrm{id}$. It is the least common multiple of its cycles lengths. For example, the order of $(132)(45)$ is $2 \cdot 3 = 6$.

## Parity of a Permutation

Every permutation of a finite set can be expressed as the product of transpositions. Although many such expressions for a given permutation may exist, either they all contain an even or an odd number of transpositions. Thus all permutations can be classified as even or odd depending on this number.

This result can be extended so as to assign a *sign*, written $\operatorname{sgn}\sigma$, to each permutation. $\operatorname{sgn}\sigma = +1$ if $\sigma$ is even and $\operatorname{sgn}\sigma = -1$ if $\sigma$ is odd. Then for two permutations $\sigma$ and $\pi$:

$$\operatorname{sgn}(\sigma\pi) = \operatorname{sgn}\sigma \cdot \operatorname{sgn}\pi$$

It follows that $\operatorname{sgn}(\sigma\sigma^{-1}) = +1.$

## Matrix Representation

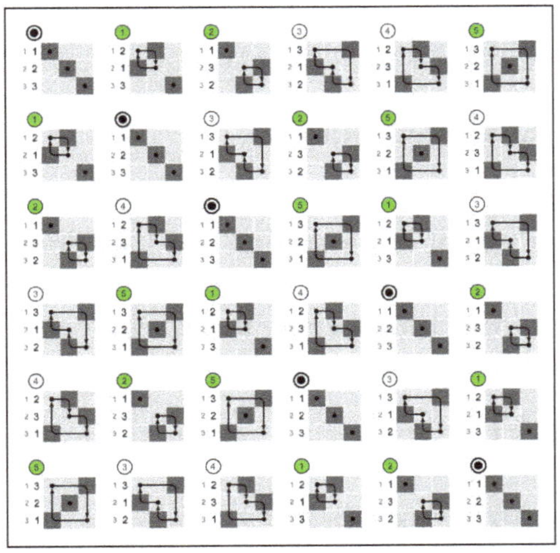

Composition of permutations corresponds to multiplication of permutation matrices.

One can represent a permutation of {1, 2, ..., $n$} as an $n \times n$ matrix. There are two natural ways to do so, but only one for which multiplications of matrices corresponds to multiplication of permutations in the same order: this is the one that associates to $\sigma$ the matrix $M$ whose entry $M_{i,j}$ is 1 if

$i = \sigma(j)$, and 0 otherwise. The resulting matrix has exactly one entry 1 in each column and in each row, and is called a *permutation matrix*. The Cayley table on the right shows these matrices for permutations of 3 elements.

## Foata's Transition Lemma

There is a relationship between the one-line and the canonical cycle notation. Consider the permutation (2)(31), in canonical cycle notation, if we erase its cycle parentheses, we obtain the permutation $(2,3,1)$ in one-line notation. Foata's transition lemma establishes the nature of this correspondence as a bijection on the set of n-permutations (to itself). Richard P. Stanley calls this correspondence the fundamental bijection.

Let $f$ be the parentheses-erasing transformation. The inverse of $q = f(p)$ is a bit less intuitive. Starting with the one-line notation $q = q_1 q_2 \cdots q_n$, the first cycle in canonical cycle notation must start with $q_1$. As long as the subsequent elements are smaller than $q_1$, we are in the same cycle. The second cycle starts at the smallest index $j$ such that $q_j > q_1$. In other words, $q_j$ is larger than everything else to its left, so it is called a *left-to-right maximum*. Every cycle in the canonical cycle notation starts with a left-to-right maximum.

For example, in the one-line notation $(3,1,2,5,4,8,9,7,6)$, 5 is the first element larger than 3, so the first cycle must be $(312)$. Then 8 is the next element larger than 5, so the second cycle is $(54)$. Since 9 is larger than 8, $(8)$ is a cycle by itself. Finally, 9 is larger than all the remaining elements to its right, so the last cycle is $(976)$.

As a first corollary, the number of n-permutations with exactly $k$ left-to-right maxima is also equal to the signless Stirling number of the first kind, $c(n,k)$. Furthermore, Foata's mapping takes an $n$-permutation with $k$-weak excedances to an $n$-permutations with $k - 1$ ascents. For example, (2)(31) = 321 has two weak excedances (at index 1 and 2), whereas $f(321) = 231$ has one ascent (at index 1; i.e., from 2 to 3).

## Permutations of Totally Ordered Sets

In some applications, the elements of the set being permuted will be compared with each other. This requires that the set $S$ has a total order so that any two elements can be compared. The set $\{1, 2, ..., n\}$ is totally ordered by the usual "≤" relation and so it is the most frequently used set in these applications, but in general, any totally ordered set will do. In these applications, the ordered arrangement view of a permutation is needed to talk about the *positions* in a permutation.

There are a number of properties that are directly related to the total ordering of $S$.

## Ascents, Descents, Runs and Excedances

An *ascent* of a permutation $\sigma$ of $n$ is any position $i < n$ where the following value is bigger than the current one. That is, if $\sigma = \sigma_1\sigma_2...\sigma_n$, then $i$ is an ascent if $\sigma_i < \sigma_{i+1}$.

For example, the permutation 3452167 has ascents (at positions) 1, 2, 5, and 6.

Similarly, a *descent* is a position $i < n$ with $\sigma_i > \sigma_{i+1}$, so every $i$ with $1 \leq i < n$ either is an ascent or is a descent of $\sigma$.

An *ascending run* of a permutation is a nonempty increasing contiguous subsequence of the permutation that cannot be extended at either end; it corresponds to a maximal sequence of successive ascents (the latter may be empty: between two successive descents there is still an ascending run of length 1). By contrast an *increasing subsequence* of a permutation is not necessarily contiguous: it is an increasing sequence of elements obtained from the permutation by omitting the values at some positions. For example, the permutation 2453167 has the ascending runs 245, 3, and 167, while it has an increasing subsequence 2367.

If a permutation has $k - 1$ descents, then it must be the union of $k$ ascending runs.

The number of permutations of $n$ with $k$ ascents is (by definition) the Eulerian number $\left\langle {n \atop k} \right\rangle$; this is also the number of permutations of $n$ with $k$ descents.

An excedance of a permutation $\sigma_1\sigma_2...\sigma_n$ in an index $j$ such that $\sigma_j > j$. If the inequality is not strict (i.e., $\sigma_j \geq j$), then $j$ is called a *weak excedance*. The number of $n$-permutations with $k$ excedances coincides with the number of $n$-permutations with $k$ descents.

## Inversions

In the 15 puzzle the goal is to get the squares in ascending order. Initial positions which have an odd number of inversions are impossible to solve.

An *inversion* of a permutation $\sigma$ is a pair $(i,j)$ of positions where the entries of a permutation are in the opposite order: $i < j$ and σ_i > σ_j. So a descent is just an inversion at two adjacent positions. For example, the permutation $\sigma$ = 23154 has three inversions: (1,3), (2,3), (4,5), for the pairs of entries (2,1), (3,1), (5,4).

Sometimes an inversion is defined as the pair of values $(\sigma_i,\sigma_j)$ itself whose order is reversed; this makes no difference for the *number* of inversions, and this pair (reversed) is also an inversion in the above sense for the inverse permutation $\sigma^{-1}$. The number of inversions is an important measure for the degree to which the entries of a permutation are out of order; it is the same for $\sigma$ and for $\sigma^{-1}$. To bring a permutation with $k$ inversions into order (i.e., transform it into the identity permutation), by successively applying (right-multiplication by) adjacent transpositions, is always possible and requires a sequence of $k$ such operations. Moreover, any reasonable choice for the adjacent transpositions will work: it suffices to choose at each step a transposition of $i$ and $i + 1$ where $i$ is a descent of the permutation as modified so far (so that the transposition will remove this particular descent, although it might create other descents). This is so because applying such a transposition reduces the number of inversions by 1; as long as this number is not zero, the permutation is not the identity, so it has at least one descent. Bubble sort and insertion sort can be interpreted as particular instances of this procedure to put a sequence into order. Incidentally this procedure proves that any permutation $\sigma$ can be written as a product of adjacent transpositions; for this one may simply reverse any sequence of such transpositions that transforms $\sigma$ into the identity. In fact, by enumerating all sequences of adjacent transpositions that would transform $\sigma$ into the identity, one obtains (after reversal) a *complete* list of all expressions of minimal length writing $\sigma$ as a product of adjacent transpositions.

The number of permutations of $n$ with $k$ inversions is expressed by a Mahonian number, it is the coefficient of $X^k$ in the expansion of the product,

$$\prod_{m=1}^{n}\sum_{i=0}^{m-1} X^i = 1(1+X)(1+X+X^2)\cdots(1+X+X^2+\cdots+X^{n-1}),$$

which is also known (with $q$ substituted for $X$) as the q-factorial $[n]_q!$. The expansion of the product appears in Necklace (combinatorics).

## Permutations in Computing

### Numbering Permutations

One way to represent permutations of $n$ is by an integer $N$ with $0 \le N < n!$, provided convenient methods are given to convert between the number and the representation of a permutation as an ordered arrangement (sequence). This gives the most compact representation of arbitrary permutations, and in computing is particularly attractive when $n$ is small enough that $N$ can be held in a machine word; for 32-bit words this means $n \le 12$, and for 64-bit words this means $n \le 20$. The conversion can be done via the intermediate form of a sequence of numbers $d_n, d_{n-1}, \ldots, d_2, d_1$, where $d_i$ is a non-negative integer less than $i$ (one may omit $d_1$, as it is always 0, but its presence makes the subsequent conversion to a permutation easier to describe). The first step then is to simply express $N$ in the *factorial number system*, which is just a particular mixed radix representation, where for numbers up to $n!$ the bases for successive digits are $n, n - 1, \ldots, 2, 1$. The second step interprets this sequence as a Lehmer code or (almost equivalently) as an inversion table.

| Rothe diagram for $\sigma = (6,3,8,1,4,9,7,2,5)$ | | | | | | | | | | |
|---|---|---|---|---|---|---|---|---|---|---|
| $i$ \ $\sigma_i$ | 1 | 2 | 3 | 4 | 5 | 6 | 7 | 8 | 9 | Lehmer code |
| 1 | × | × | × | × | × | • | | | | $d_9 = 5$ |
| 2 | × | × | • | | | | | | | $d_8 = 2$ |
| 3 | × | × | | × | × | | × | • | | $d_7 = 5$ |

| 4 | • | | | | | | | | | $d_6 = 0$ |
|---|---|---|---|---|---|---|---|---|---|---|
| 5 | | × | | • | | | | | | $d_5 = 1$ |
| 6 | | × | | | × | | × | | • | $d_4 = 3$ |
| 7 | | × | | | × | | • | | | $d_3 = 2$ |
| 8 | | • | | | | | | | | $d_2 = 0$ |
| 9 | | | | | • | | | | | $d_1 = 0$ |
| Inversion table | 3 | 6 | 1 | 2 | 4 | 0 | 2 | 0 | 0 | |

In the Lehmer code for a permutation $\sigma$, the number $d_n$ represents the choice made for the first term $\sigma_1$, the number $d_{n-1}$ represents the choice made for the second term $\sigma_2$ among the remaining $n - 1$ elements of the set, and so forth. More precisely, each $d_{n+1-i}$ gives the number of *remaining* elements strictly less than the term $\sigma_i$. Since those remaining elements are bound to turn up as some later term $\sigma_j$, the digit $d_{n+1-i}$ counts the *inversions* $(i,j)$ involving $i$ as smaller index (the number of values $j$ for which $i < j$ and $\sigma_i > \sigma_j$). The inversion table for $\sigma$ is quite similar, but here $d_{n+1-k}$ counts the number of inversions $(i,j)$ where $k = \sigma_j$ occurs as the smaller of the two values appearing in inverted order. Both encodings can be visualized by an $n$ by $n$ Rothe diagram (named after Heinrich August Rothe) in which dots at $(i,\sigma_i)$ mark the entries of the permutation, and a cross at $(i,\sigma_j)$ marks the inversion $(i,j)$; by the definition of inversions a cross appears in any square that comes both before the dot $(j,\sigma_j)$ in its column, and before the dot $(i,\sigma_i)$ in its row. The Lehmer code lists the numbers of crosses in successive rows, while the inversion table lists the numbers of crosses in successive columns; it is just the Lehmer code for the inverse permutation, and vice versa.

To effectively convert a Lehmer code $d_n, d_{n-1}, ..., d_2, d_1$ into a permutation of an ordered set $S$, one can start with a list of the elements of $S$ in increasing order, and for $i$ increasing from 1 to $n$ set $\sigma_i$ to the element in the list that is preceded by $d_{n+1-i}$ other ones, and remove that element from the list. To convert an inversion table $d_n, d_{n-1}, ..., d_2, d_1$ into the corresponding permutation, one can traverse the numbers from $d_1$ to $d_n$ while inserting the elements of $S$ from largest to smallest into an initially empty sequence; at the step using the number $d$ from the inversion table, the element from $S$ inserted into the sequence at the point where it is preceded by $d$ elements already present. Alternatively one could process the numbers from the inversion table and the elements of $S$ both in the opposite order, starting with a row of $n$ empty slots, and at each step place the element from $S$ into the empty slot that is preceded by $d$ other empty slots.

Converting successive natural numbers to the factorial number system produces those sequences in lexicographic order (as is the case with any mixed radix number system), and further converting them to permutations preserves the lexicographic ordering, provided the Lehmer code interpretation is used (using inversion tables, one gets a different ordering, where one starts by comparing permutations by the *place* of their entries 1 rather than by the value of their first entries). The sum of the numbers in the factorial number system representation gives the number of inversions of the permutation, and the parity of that sum gives the signature of the permutation. Moreover, the positions of the zeroes in the inversion table give the values of left-to-right maxima of the permutation (in the example 6, 8, 9) while the positions of the zeroes in the Lehmer code are the positions of the right-to-left minima (in the example positions the 4, 8, 9 of the values 1, 2, 5); this allows computing the distribution of such extrema among all permutations. A permutation with Lehmer code $d_n, d_{n-1}, ..., d_2, d_1$ has an ascent $n - i$ if and only if $d_i \geq d_{i+1}$.

## Algorithms to Generate Permutations

In computing it may be required to generate permutations of a given sequence of values. The methods best adapted to do this depend on whether one wants some randomly chosen permutations, or all permutations, and in the latter case if a specific ordering is required. Another question is whether possible equality among entries in the given sequence is to be taken into account; if so, one should only generate distinct multiset permutations of the sequence.

An obvious way to generate permutations of $n$ is to generate values for the Lehmer code (possibly using the factorial number system representation of integers up to $n!$), and convert those into the corresponding permutations. However, the latter step, while straightforward, is hard to implement efficiently, because it requires $n$ operations each of selection from a sequence and deletion from it, at an arbitrary position; of the obvious representations of the sequence as an array or a linked list, both require (for different reasons) about $n^2/4$ operations to perform the conversion. With $n$ likely to be rather small (especially if generation of all permutations is needed) that is not too much of a problem, but it turns out that both for random and for systematic generation there are simple alternatives that do considerably better. For this reason it does not seem useful, although certainly possible, to employ a special data structure that would allow performing the conversion from Lehmer code to permutation in $O(n \log n)$ time.

## Random Generation of Permutations

For generating random permutations of a given sequence of $n$ values, it makes no difference whether one applies a randomly selected permutation of $n$ to the sequence, or chooses a random element from the set of distinct (multiset) permutations of the sequence. This is because, even though in case of repeated values there can be many distinct permutations of $n$ that result in the same permuted sequence, the number of such permutations is the same for each possible result. Unlike for systematic generation, which becomes unfeasible for large $n$ due to the growth of the number $n!$, there is no reason to assume that $n$ will be small for random generation.

The basic idea to generate a random permutation is to generate at random one of the $n!$ sequences of integers $d_1,d_2,\ldots,d_n$ satisfying $0 \le d_i < i$ (since $d_1$ is always zero it may be omitted) and to convert it to a permutation through a bijective correspondence. For the latter correspondence one could interpret the (reverse) sequence as a Lehmer code, and this gives a generation method first published in 1938 by Ronald Fisher and Frank Yates. While at the time computer implementation was not an issue, this method suffers from the difficulty sketched above to convert from Lehmer code to permutation efficiently. This can be remedied by using a different bijective correspondence: after using $d_i$ to select an element among $i$ remaining elements of the sequence (for decreasing values of $i$), rather than removing the element and compacting the sequence by shifting down further elements one place, one swaps the element with the final remaining element. Thus the elements remaining for selection form a consecutive range at each point in time, even though they may not occur in the same order as they did in the original sequence. The mapping from sequence of integers to permutations is somewhat complicated, but it can be seen to produce each permutation in exactly one way, by an immediate induction. When the selected element happens to be the final remaining element, the swap operation can be omitted. This does not occur sufficiently often to warrant testing for the condition, but the final element must be included among the candidates of the selection, to guarantee that all permutations can be generated.

The resulting algorithm for generating a random permutation of `a[0], a[1], ..., a[n - 1]` can be described as follows in pseudocode:

```
for i from n downto 2
do d_i ← random element of { 0, ..., i - 1 }
   swap a[d_i] and a[i - 1]
```

This can be combined with the initialization of the array $a[i] = i$ as follows:

```
for i from 0 to n-1
do d_{i+1} ← random element of { 0, ..., i }
      a[i] ← a[d_{i+1}]
      a[d_{i+1}] ← i
```

If $d_{i+1} = i$, the first assignment will copy an uninitialized value, but the second will overwrite it with the correct value $i$.

However, Fisher-Yates is not the fastest algorithm for generating a permutation, because Fisher-Yates is essentially a sequential algorithm and "divide and conquer" procedures can achieve the same result in parallel.

## Generation in Lexicographic Order

There are many ways to systematically generate all permutations of a given sequence. One classic, simple, and flexible algorithm is based upon finding the next permutation in lexicographic ordering, if it exists. It can handle repeated values, for which case it generates each distinct multiset permutation once. Even for ordinary permutations it is significantly more efficient than generating values for the Lehmer code in lexicographic order (possibly using the factorial number system) and converting those to permutations. It begins by sorting the sequence in (weakly) increasing order (which gives its lexicographically minimal permutation), and then repeats advancing to the next permutation as long as one is found.

The following algorithm generates the next permutation lexicographically after a given permutation. It changes the given permutation in-place.

1. Find the largest index $k$ such that $a[k] < a[k + 1]$. If no such index exists, the permutation is the last permutation.
2. Find the largest index $l$ greater than k such that $a[k] < a[l]$.
3. Swap the value of $a[k]$ with that of $a[l]$.
4. Reverse the sequence from $a[k + 1]$ up to and including the final element $a[n]$.

For example, given the sequence [1, 2, 3, 4] (which is in increasing order), and given that the index is zero-based, the steps are as follows:

1. Index $k = 2$, because 3 is placed at an index that satisfies condition of being the largest index that is still less than $a[k + 1]$ which is 4.

2. Index $l = 3$, because 4 is the only value in the sequence that is greater than 3 in order to satisfy the condition $a[k] < a[l]$.

3. The values of $a$ and $a$ are swapped to form the new sequence [1,2,4,3].

4. The sequence after $k$-index $a$ to the final element is reversed. Because only one value lies after this index (the 3), the sequence remains unchanged in this instance. Thus the lexicographic successor of the initial state is permuted: [1,2,4,3].

Following this algorithm, the next lexicographic permutation will be [1,3,2,4], and the 24th permutation will be [4,3,2,1] at which point $a[k] < a[k + 1]$ does not exist, indicating that this is the last permutation.

This method uses about 3 comparisons and 1.5 swaps per permutation, amortized over the whole sequence, not counting the initial sort.

## Generation with Minimal Changes

An alternative to the above algorithm, the Steinhaus–Johnson–Trotter algorithm, generates an ordering on all the permutations of a given sequence with the property that any two consecutive permutations in its output differ by swapping two adjacent values. This ordering on the permutations was known to 17th-century English bell ringers, among whom it was known as "plain changes". One advantage of this method is that the small amount of change from one permutation to the next allows the method to be implemented in constant time per permutation. The same can also easily generate the subset of even permutations, again in constant time per permutation, by skipping every other output permutation.

An alternative to Steinhaus–Johnson–Trotter is Heap's algorithm, the fastest algorithm of generating permutations in applications.

The following figure shows the output of all three aforementioned algorithms for generating all permutations of length $n = 4$, and of six additional algorithms described in the literature.

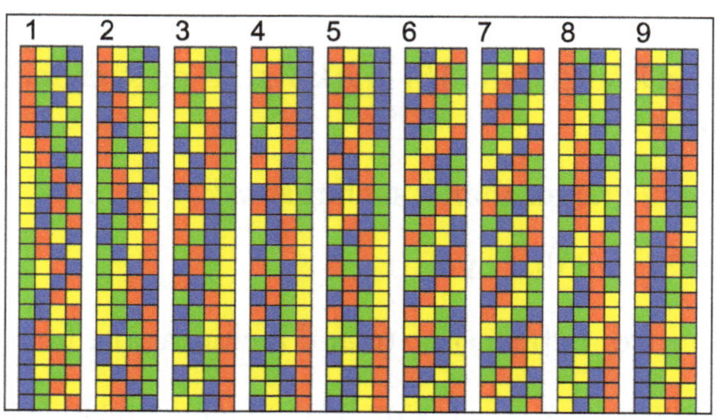


Ordering of all permutations of length $n = 4$, generated by different algorithms. The permutations are color-coded, where 1 = red, 2 = yellow, 3 = green, 4 = blue.

1: Lexicographic ordering; 2: Steinhaus–Johnson–Trotter algorithm; 3: Heap's algorithm; 4: Ehrlich's star-transposition algorithm: in each step, the first entry of the permutation is exchanged with a later entry; 5: Zaks' prefix reversal algorithm: in each step, a prefix of the current permutation is reversed to obtain the next permutation; 6: Sawada-Williams' algorithm: each permutation differs from the previous one either by a cyclic left-shift by one position, or an exchange of the first two entries; 7: Corbett's algorithm: each permutation differs from the previous one by a cyclic left-shift of some prefix by one position; 8: Single-track ordering: each column is a cyclic shift of the other columns; 9: Single-track Gray code: each column is a cyclic shift of the other columns, plus any two consecutive permutations differ only in one or two transpositions.

## Meandric Permutations

Meandric systems give rise to *meandric permutations*, a special subset of *alternate permutations*. An alternate permutation of the set $\{1, 2, ..., 2n\}$ is a cyclic permutation (with no fixed points) such that the digits in the cyclic notation form alternate between odd and even integers. Meandric permutations are useful in the analysis of RNA secondary structure. Not all alternate permutations are meandric. A modification of Heap's algorithm has been used to generate all alternate permutations of order $n$ (that is, of length $2n$) without generating all $(2n)!$ permutations. Generation of these alternate permutations is needed before they are analyzed to determine if they are meandric or not.

The algorithm is recursive. The following table exhibits a step in the procedure. In the previous step, all alternate permutations of length 5 have been generated. Three copies of each of these have a "6" added to the right end, and then a different transposition involving this last entry and a previous entry in an even position is applied (including the identity; i.e., no transposition).

| Previous sets | Transposition of digits | Alternate permutations |
|---|---|---|
| 1-2-3-4-5-6 | | 1-2-3-4-5-6 |
| | 4, 6 | 1-2-3-6-5-4 |
| | 2, 6 | 1-6-3-4-5-2 |
| 1-2-5-4-3-6 | | 1-2-5-4-3-6 |
| | 4, 6 | 1-2-5-6-3-4 |
| | 2, 6 | 1-6-5-4-3-2 |
| 1-4-3-2-5-6 | | 1-4-3-2-5-6 |
| | 2, 6 | 1-4-3-6-5-2 |
| | 4, 6 | 1-6-3-2-5-4 |
| 1-4-5-2-3-6 | | 1-4-5-2-3-6 |
| | 2, 6 | 1-4-5-6-3-2 |
| | 4, 6 | 1-6-5-2-3-4 |

## Applications

Permutations are used in the interleaver component of the error detection and correction algorithms, such as turbo codes, for example 3GPP Long Term Evolution mobile telecommunication standard uses these ideas. Such applications raise the question of fast generation of permutations satisfying certain desirable properties. One of the methods is based on the permutation polynomials. Also as a base for optimal hashing in Unique Permutation Hashing.

## Cyclic Permutation

In mathematics, and in particular in group theory, a cyclic permutation (or cycle) is a permutation of the elements of some set $X$ which maps the elements of some subset $S$ of $X$ to each other in a cyclic fashion, while fixing (that is, mapping to themselves) all other elements of $X$. If $S$ has $k$ elements, the cycle is called a k-cycle. Cycles are often denoted by the list of their elements enclosed with parentheses, in the order to which they are permuted.

For example, given $X = \{1, 2, 3, 4\}$, the permutation (1, 3, 2, 4) that sends 1 to 3, 3 to 2, 2 to 4 and 4 to 1 (so $S = X$) is a 4-cycle, and the permutation (1, 3, 2) that sends 1 to 3, 3 to 2, 2 to 1 and 4 to 4 (so $S = \{1, 2, 3\}$ and 4 is a fixed element) is a 3-cycle. On the other hand, the permutation that sends 1 to 3, 3 to 1, 2 to 4 and 4 to 2 is not a cyclic permutation because it separately permutes the pairs {1, 3} and {2, 4}.

The set $S$ is called the orbit of the cycle. Every permutation on finitely many elements can be decomposed into cycles on disjoint orbits.

The cyclic parts of a permutation are cycles, thus the second example is composed of a 3-cycle and a 1-cycle (or *fixed point*) and the third is composed of two 2-cycles, and denoted (1, 3) (2, 4).

A permutation is called a cyclic permutation if and only if it has a single nontrivial cycle (a cycle of length > 1).

For example, the permutation, written in two-line (in two ways) and also cycle notations,

$$\begin{pmatrix} 1 & 2 & 3 & 4 & 5 & 6 & 7 & 8 \\ 4 & 2 & 7 & 6 & 5 & 8 & 1 & 3 \end{pmatrix} = \begin{pmatrix} 1 & 4 & 6 & 8 & 3 & 7 & 2 & 5 \\ 4 & 6 & 8 & 3 & 7 & 1 & 2 & 5 \end{pmatrix} = (146837)(2)(5),$$

is a six-cycle; its cycle diagram is shown below.

Some authors restrict the definition to only those permutations which consist of one nontrivial cycle (that is, no fixed points allowed).

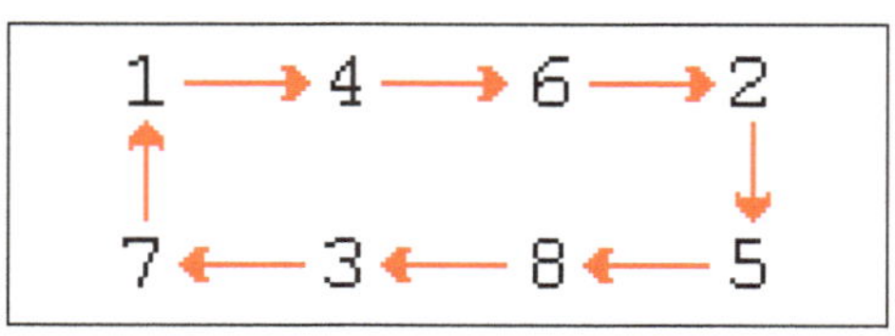


A cyclic permutation with no trivial cycles; an 8-cycle.

For example, the permutation,

$$\begin{pmatrix} 1 & 2 & 3 & 4 & 5 & 6 & 7 & 8 \\ 4 & 5 & 7 & 6 & 8 & 2 & 1 & 3 \end{pmatrix} = \begin{pmatrix} 1 & 4 & 6 & 2 & 5 & 8 & 3 & 7 \\ 4 & 6 & 2 & 5 & 8 & 3 & 7 & 1 \end{pmatrix} = (14625837)$$

is a cyclic permutation under this more restrictive definition, while the preceding example is not.

More formally, a permutation $\sigma$ of a set $X$, viewed as a bijective function $\sigma : X \to X$, is called a cycle if the action on $X$ of the subgroup generated by has at most one orbit with more than a single element. This notion is most commonly used when $X$ is a finite set; then of course the largest orbit, $S$, is also finite. Let $s_0$ be any element of $S$, and put $s_i = \sigma^i(s_0)$ for any $i \in \mathbf{Z}$. If $S$ is finite, there is a minimal number $k \geq 1$ for which $s_k = s_0$. Then $S = \{s_0, s_1, \ldots, s_{k-1}\}$, and $\sigma$ is the permutation defined by

$$\sigma(s_i) = s_{i+1} \text{ for } 0 \leq i < k$$

and $\sigma(x) = x$ for any element of X\S. The elements not fixed by $\sigma$ can be pictured as:

$$s_0 \mapsto s_1 \mapsto s_2 \mapsto \cdots \mapsto s_{k-1} \mapsto s_k = s_0.$$

A cycle can be written using the compact cycle notation $\sigma = (s_0\ s_1 \ldots s_{k-1})$ (there are no commas between elements in this notation, to avoid confusion with a $k$-tuple). The *length* of a cycle is the number of elements of its largest orbit. A cycle of length $k$ is also called a $k$-cycle.

The orbit of a 1-cycle is called a *fixed point* of the permutation, but as a permutation every 1-cycle is the identity permutation. When cycle notation is used, the 1-cycles are often suppressed when no confusion will result.

## Basic Properties

One of the basic results on symmetric groups says that any permutation can be expressed as the product of disjoint cycles (more precisely: cycles with disjoint orbits); such cycles commute with each other, and the expression of the permutation is unique up to the order of the cycles. The multiset of lengths of the cycles in this expression (the cycle type) is therefore uniquely determined by the permutation, and both the signature and the conjugacy class of the permutation in the symmetric group are determined by it.

The number of $k$-cycles in the symmetric group $S_n$ is given, for $1 \leq k \leq n$, by the following equivalent formulas:

$$\binom{n}{k}(k-1)! = \frac{n(n-1)\cdots(n-k+1)}{k} = \frac{n!}{(n-k)!k}$$

A $k$-cycle has signature $(-1)^{k-1}$.

The inverse of a cycle $\sigma = (s_0\ s_1 \ldots s_{k-1})$ is given by reversing the order of the entries: $\sigma^{-1} = (s_{k-1} \ldots s_1\ s_0)$. In particular, since $(a\,b) = (b\,a)$, every two-cycle is its own inverse. Since disjoint cycles commute, the inverse of a product of disjoint cycles is the result of reversing each of the cycles separately.

## Transpositions

A cycle with only two elements is called a transposition. For example, the permutation $\pi = \begin{pmatrix} 1 & 2 & 3 & 4 \\ 1 & 4 & 3 & 2 \end{pmatrix}$ that swaps 2 and 4.

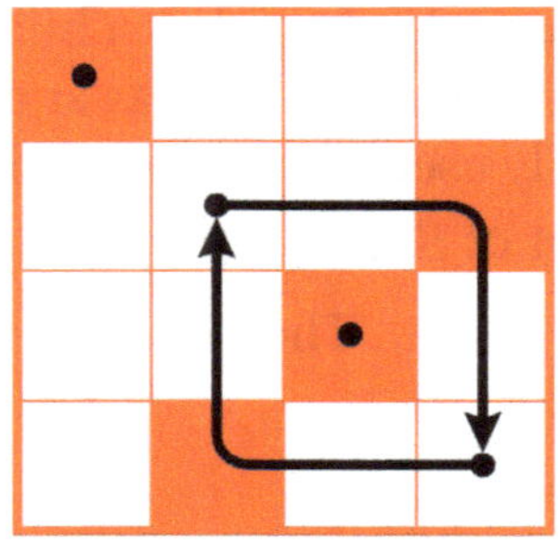

Matrix of $\pi$

## Properties

Any permutation can be expressed as the composition (product) of transpositions—formally, they are generators for the group. In fact, when the set being permuted is {1, 2, ..., $n$} for some integer $n$, then any permutation can be expressed as a product of adjacent transpositions $(12),(23),(34)$, and so on. This follows because an arbitrary transposition can be expressed as the product of adjacent transpositions. Concretely, one can express the transposition $(k\ l)$ where $k < l$ by moving $k$ to $l$ one step at a time, then moving $l$ back to where $k$ was, which interchanges these two and makes no other changes:

$$(k\ l) = (k\ k+1)\cdot(k+1\ k+2)\cdots(l-1\ l)\cdot(l-2\ l-1)\cdots(k\ k+1).$$

The decomposition of a permutation into a product of transpositions is obtained for example by writing the permutation as a product of disjoint cycles, and then splitting iteratively each of the cycles of length 3 and longer into a product of a transposition and a cycle of length one less:

$$(abcd\ldots yz) = (ab)\cdot(bcd\ldots yz).$$

This means the initial request is to move $a$ to $b$ ,$b$ to $c$,   to $z$, and finally $z$, to $a$ Instead one may roll the elements keeping $a$ where it is by executing the right factor first. This has moved $z$, to the position of $b$ so after the first permutation, the elements $a$ and $z$ are not yet at their final positions. The transposition $(ab)$, executed thereafter, then addresses $z$ by the index of $b$ to swap what initially were $a$ and $z$ .

In fact, the symmetric group is a Coxeter group, meaning that it is generated by elements of order 2 (the adjacent transpositions), and all relations are of a certain form.

One of the main results on symmetric groups states that either all of the decompositions of a given permutation into transpositions have an even number of transpositions, or they all have an odd number of transpositions. This permits the parity of a permutation to be a well-defined concept.

# Combination

In mathematics, a combination is a selection of items from a collection, such that (unlike permutations) the order of selection does not matter. For example, given three fruits, say an apple, an orange and a pear, there are three combinations of two that can be drawn from this set: an apple and a pear; an apple and an orange; or a pear and an orange. More formally, a $k$-combination of a set $S$ is a subset of $k$ distinct elements of $S$. If the set has $n$ elements, the number of $k$-combinations is equal to the binomial coefficient,

$$\binom{n}{k} = \frac{n(n-1)\cdots(n-k+1)}{k(k-1)\cdots 1},$$

which can be written using factorials as $\frac{n!}{k!(n-k)!}$ whenever $k \leq n$, and which is zero when $k > n$. The set of all $k$-combinations of a set $S$ is often denoted by $\binom{S}{k}$.

Combinations refer to the combination of $n$ things taken $k$ at a time without repetition. To refer to combinations in which repetition is allowed, the terms $k$-selection, $k$-multiset, or $k$-combination with repetition are often used. If, in the above example, it were possible to have two of any one kind of fruit there would be 3 more 2-selections: one with two apples, one with two oranges, and one with two pears.

Although the set of three fruits was small enough to write a complete list of combinations, with large sets this becomes impractical. For example, a poker hand can be described as a 5-combination ($k$ = 5) of cards from a 52 card deck ($n$ = 52). The 5 cards of the hand are all distinct, and the order of cards in the hand does not matter. There are 2,598,960 such combinations, and the chance of drawing any one hand at random is 1 / 2,598,960.

## Number of *k*-combinations

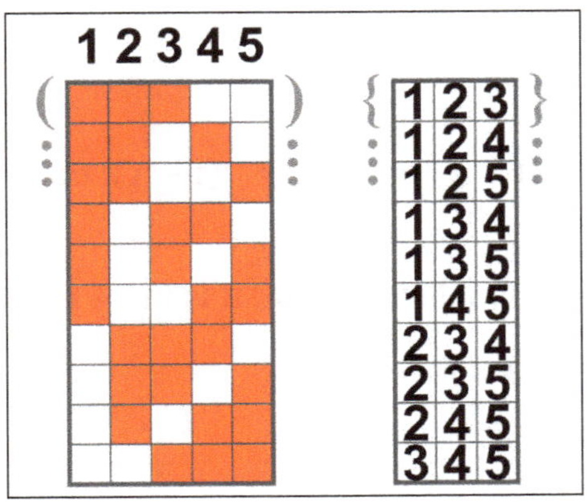


3-element subsets of a 5-element set.

The number of $k$-combinations from a given set $S$ of $n$ elements is often denoted in elementary combinatorics texts by $C(n,k)$, or by a variation such as $C_k^n$, ${}_nC_k$, ${}^nC_k$, $C_{n,k}$ or even $C_n^k$ (the latter form was standard in French, Romanian, Russian, Chinese and Polish texts. The same number however occurs in many other mathematical contexts, where it is denoted by $\binom{n}{k}$ (often read as "$n$ choose $k$"); notably it occurs as a coefficient in the binomial formula, hence its name binomial coefficient. One can define $\binom{n}{k}$ for all natural numbers $k$ at once by the relation,

$$(1+X)^n = \sum_{k\geq 0}\binom{n}{k}X^k,$$

from which it is clear that,

$$\binom{n}{0} = \binom{n}{n} = 1,$$

and further,

$$\binom{n}{k} = 0 \text{ for } k > n.$$

To see that these coefficients count $k$-combinations from $S$, one can first consider a collection of $n$ distinct variables $X_s$ labeled by the elements $s$ of $S$, and expand the product over all elements of $S$:

$$\prod_{s\in S}(1+X_s)$$

It has $2^n$ distinct terms corresponding to all the subsets of $S$, each subset giving the product of the corresponding variables $X_s$. Now setting all of the $X_s$ equal to the unlabeled variable $X$, so that the product becomes $(1 + X)^n$, the term for each $k$-combination from $S$ becomes $X^k$, so that the coefficient of that power in the result equals the number of such $k$-combinations.

Binomial coefficients can be computed explicitly in various ways. To get all of them for the expansions up to $(1 + X)^n$, one can use (in addition to the basic cases already given) the recursion relation,

$$\binom{n}{k} = \binom{n-1}{k-1} + \binom{n-1}{k},$$

for $0 < k < n$, which follows from $(1 + X)^n = (1 + X)^{n-1}(1 + X)$; this leads to the construction of Pascal's triangle.

For determining an individual binomial coefficient, it is more practical to use the formula:

$$\binom{n}{k} = \frac{n(n-1)(n-2)\cdots(n-k+1)}{k!}.$$

The numerator gives the number of $k$-permutations of $n$, i.e., of sequences of $k$ distinct elements of $S$, while the denominator gives the number of such $k$-permutations that give the same $k$-combination when the order is ignored.

When $k$ exceeds $n/2$, the above formula contains factors common to the numerator and the denominator, and canceling them out gives the relation:

$$\binom{n}{k} = \binom{n}{n-k}$$

for $0 \le k \le n$. This expresses a symmetry that is evident from the binomial formula, and can also be understood in terms of $k$-combinations by taking the complement of such a combination, which is an $(n - k)$-combination.

Finally there is a formula which exhibits this symmetry directly, and has the merit of being easy to remember:

$$\binom{n}{k} = \frac{n!}{k!(n-k)!},$$

where $n!$ denotes the factorial of $n$. It is obtained from the previous formula by multiplying denominator and numerator by $(n - k)!$, so it is certainly inferior as a method of computation to that formula.

The last formula can be understood directly, by considering the $n!$ permutations of all the elements of $S$. Each such permutation gives a $k$-combination by selecting its first $k$ elements. There are many duplicate selections: any combined permutation of the first $k$ elements among each other, and of the final $(n - k)$ elements among each other produces the same combination; this explains the division in the formula.

From the above formulas follow relations between adjacent numbers in Pascal's triangle in all three directions:

$$\binom{n}{k} = \begin{cases} \binom{n}{k-1}\frac{n-k+1}{k} & \text{if } k > 0 \\ \binom{n-1}{k}\frac{n}{n-k} & \text{if } k < n \\ \binom{n-1}{k-1}\frac{n}{k} & \text{if } n, k > 0 \end{cases}.$$

Together with the basic cases $\binom{n}{0} = 1 = \binom{n}{n}$, these allow successive computation of respectively all numbers of combinations from the same set (a row in Pascal's triangle), of $k$-combinations of sets of growing sizes, and of combinations with a complement of fixed size $n - k$.

## Example of Counting Combinations

As a specific example, one can compute the number of five-card hands possible from a standard fifty-two card deck as:

$$\binom{52}{5} = \frac{52\times51\times50\times49\times48}{5\times4\times3\times2\times1} = \frac{311,875,200}{120} = 2,598,960.$$

Alternatively one may use the formula in terms of factorials and cancel the factors in the numerator against parts of the factors in the denominator, after which only multiplication of the remaining factors is required.

$$
\begin{aligned}
\binom{52}{5} &= \frac{52!}{5!47!} \\
&= \frac{52\times51\times50\times49\times48\times\cancel{47}!}{5\times4\times3\times2\times\cancel{1}\times\cancel{47}!} \\
&= \frac{52\times51\times50\times49\times48}{5\times4\times3\times2} \\
&= \frac{(26\times\cancel{2})\times(17\times\cancel{3})\times(10\times\cancel{5})\times49\times(12\times\cancel{4})}{\cancel{5}\times\cancel{4}\times\cancel{3}\times\cancel{2}} \\
&= 26\times17\times10\times49\times12 \\
&= 2{,}598{,}960.
\end{aligned}
$$

Another alternative computation, equivalent to the first, is based on writing,

$$\binom{n}{k} = \frac{(n-0)}{1}\times\frac{(n-1)}{2}\times\frac{(n-2)}{3}\times\cdots\times\frac{(n-(k-1))}{k},$$

which gives,

$$\binom{52}{5} = \frac{52}{1}\times\frac{51}{2}\times\frac{50}{3}\times\frac{49}{4}\times\frac{48}{5} = 2{,}598{,}960$$

When evaluated in the following order, $52 \div 1 \times 51 \div 2 \times 50 \div 3 \times 49 \div 4 \times 48 \div 5$, this can be computed using only integer arithmetic. The reason is that when each division occurs, the intermediate result that is produced is itself a binomial coefficient, so no remainders ever occur.

Using the symmetric formula in terms of factorials without performing simplifications gives a rather extensive calculation:

$$
\begin{aligned}
\binom{52}{5} &= \frac{n!}{k!(n-k)!} = \frac{52!}{5!(52-5)!} = \frac{52!}{5!47!} \\
&= \frac{80{,}658{,}175{,}170{,}943{,}878{,}571{,}660{,}636{,}856{,}403{,}766{,}975{,}289{,}505{,}440{,}883{,}277{,}824{,}000{,}000{,}000{,}000}{120\times258{,}623{,}241{,}511{,}168{,}180{,}642{,}964{,}355{,}153{,}611{,}979{,}969{,}197{,}632{,}389{,}120{,}000{,}000{,}000} \\
&= 2{,}598{,}960.
\end{aligned}
$$

## Enumerating *k*-combinations

One can enumerate all $k$-combinations of a given set $S$ of $n$ elements in some fixed order, which establishes a bijection from an interval of $\binom{n}{k}$ integers with the set of those $k$-combinations. Assuming $S$ is itself ordered, for instance $S = \{ 1, 2, \ldots, n \}$, there are two natural possibilities for ordering its $k$-combinations: by comparing their smallest elements first (as in the illustrations above) or by comparing their largest elements first. The latter option has the advantage that adding a new largest element to $S$ will not change the initial part of the enumeration, but just add the new

$k$-combinations of the larger set after the previous ones. Repeating this process, the enumeration can be extended indefinitely with $k$-combinations of ever larger sets. If moreover the intervals of the integers are taken to start at 0, then the $k$-combination at a given place $i$ in the enumeration can be computed easily from $i$, and the bijection so obtained is known as the combinatorial number system. It is also known as "rank"/"ranking" and "unranking" in computational mathematics.

There are many ways to enumerate $k$ combinations. One way is to visit all the binary numbers less than $2^n$. Choose those numbers having $k$ nonzero bits, although this is very inefficient even for small $n$ (e.g. $n = 20$ would require visiting about one million numbers while the maximum number of allowed $k$ combinations is about 186 thousand for $k = 10$). The positions of these 1 bits in such a number is a specific $k$-combination of the set $\{ 1, ..., n \}$. Another simple, faster way is to track $k$ index numbers of the elements selected, starting with $\{0 .. k-1\}$ (zero-based) or $\{1 .. k\}$ (one-based) as the first allowed $k$-combination and then repeatedly moving to the next allowed $k$-combination by incrementing the last index number if it is lower than $n$-1 (zero-based) or $n$ (one-based) or the last index number $x$ that is less than the index number following it minus one if such an index exists and resetting the index numbers after $x$ to $\{x+1, x+2, ...\}$.

## Number of Combinations with Repetition

A $k$-combination with repetitions, or $k$-multicombination, or multisubset of size $k$ from a set $S$ is given by a sequence of $k$ not necessarily distinct elements of $S$, where order is not taken into account: two sequences define the same multiset if one can be obtained from the other by permuting the terms. In other words, the number of ways to sample $k$ elements from a set of $n$ elements allowing for duplicates (i.e., with replacement) but disregarding different orderings (e.g. {2,1,2} = {1,2,2}). Associate an index to each element of $S$ and think of the elements of $S$ as *types* of objects, then we can let $x_i$ denote the number of elements of type $i$ in a multisubset. The number of multisubsets of size $k$ is then the number of nonnegative integer solutions of the Diophantine equation:

$$x_1 + x_2 + \ldots + x_n = k.$$

If $S$ has $n$ elements, the number of such $k$-multisubsets is denoted by,

$$\left(\!\!\binom{n}{k}\!\!\right),$$

a notation that is analogous to the binomial coefficient which counts $k$-subsets. This expression, $n$ multichoose $k$, can also be given in terms of binomial coefficients:

$$\left(\!\!\binom{n}{k}\!\!\right) = \binom{n+k-1}{k}.$$

This relationship can be easily proved using a representation known as stars and bars.

A solution of the above Diophantine equation can be represented by $x_1$ *stars*, a separator (a *bar*), then $x_2$ more stars, another separator, and so on. The total number of stars in this representation is $k$ and the number of bars is $n$ - 1 (since no separator is needed at the very end). Thus, a string of

$k + n - 1$ symbols (stars and bars) corresponds to a solution if there are $k$ stars in the string. Any solution can be represented by choosing $k$ out of $k + n - 1$ positions to place stars and filling the remaining positions with bars. For example, the solution $x_1 = 3, x_2 = 2, x_3 = 0, x_4 = 5$ of the equation $x_1 + x_2 + x_3 + x_4 = 10$ can be represented by:

★★★|★★||★★★★★

The number of such strings is the number of ways to place 10 stars in 13 positions, $\binom{13}{10} = \binom{13}{3} = 286$ which is the number of 10-multisubsets of a set with 4 elements.

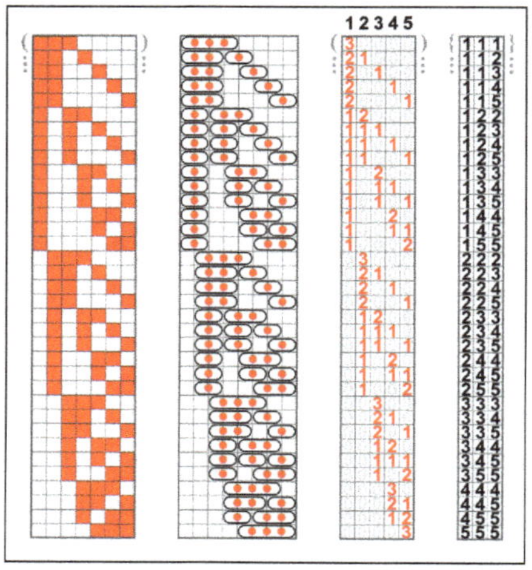

Bijection between 3-subsets of a 7-set (left) and 3-multisets with elements from a 5-set (right). This illustrates that $\binom{7}{3} = \left(\!\binom{5}{3}\!\right)$.

As with binomial coefficients, there are several relationships between these multichoose expressions. For example, for $n \geq 1, k \geq 0$,

$$\left(\!\binom{n}{k}\!\right) = \left(\!\binom{k+1}{n-1}\!\right)$$

This identity follows from interchanging the stars and bars in the above representation.

## Example of Counting Multisubsets

For example, if you have four types of donuts ($n = 4$) on a menu to choose from and you want three donuts ($k = 3$), the number of ways to choose the donuts with repetition can be calculated as:

$$\left(\!\binom{4}{3}\!\right) = \binom{4+3-1}{3} = \binom{6}{3} = \frac{6 \times 5 \times 4}{3 \times 2 \times 1} = 20.$$

This result can be verified by listing all the 3-multisubsets of the set $S = \{1,2,3,4\}$. This is displayed in the following table. The second column shows the nonnegative integer solutions $[x_1, x_2, x_3, x_4]$ of the equation $x_1 + x_2 + x_3 + x_4 = 3$ and the last column gives the stars and bars representation of the solutions.

| No. | 3-Multiset | Eq. Solution | Stars and Bars |
|---|---|---|---|
| 1 | {1,1,1} | [3,0,0,0] | ★★★\|\|\| |
| 2 | {1,1,2} | [2,1,0,0] | ★★\|★\|\| |
| 3 | {1,1,3} | [2,0,1,0] | ★★\|\|★\| |
| 4 | {1,1,4} | [2,0,0,1] | ★★\|\|\|★ |
| 5 | {1,2,2} | [1,2,0,0] | ★\|★★\|\| |
| 6 | {1,2,3} | [1,1,1,0] | ★\|★\|★\| |
| 7 | {1,2,4} | [1,1,0,1] | ★\|★\|\|★ |
| 8 | {1,3,3} | [1,0,2,0] | ★\|\|★★\| |
| 9 | {1,3,4} | [1,0,1,1] | ★\|\|★\|★ |
| 10 | {1,4,4} | [1,0,0,2] | ★\|\|\|★★ |
| 11 | {2,2,2} | [0,3,0,0] | \|★★★\|\| |
| 12 | {2,2,3} | [0,2,1,0] | \|★★\|★\| |
| 13 | {2,2,4} | [0,2,0,1] | \|★★\|\|★ |
| 14 | {2,3,3} | [0,1,2,0] | \|★\|★★\| |
| 15 | {2,3,4} | [0,1,1,1] | \|★\|★\|★ |
| 16 | {2,4,4} | [0,1,0,2] | \|★\|\|★★ |
| 17 | {3,3,3} | [0,0,3,0] | \|\|★★★\| |
| 18 | {3,3,4} | [0,0,2,1] | \|\|★★\|★ |
| 19 | {3,4,4} | [0,0,1,2] | \|\|★\|★★ |
| 20 | {4,4,4} | [0,0,0,3] | \|\|\|★★★ |

## Number of *k*-combinations for all *k*

The number of *k*-combinations for all *k* is the number of subsets of a set of *n* elements. There are several ways to see that this number is $2^n$. In terms of combinations, $\sum_{0 \le k \le n} \binom{n}{k} = 2^n$, which is the sum of the *n*th row (counting from 0) of the binomial coefficients in Pascal's triangle. These combinations (subsets) are enumerated by the 1 digits of the set of base 2 numbers counting from 0 to $2^n - 1$, where each digit position is an item from the set of *n*.

Given 3 cards numbered 1 to 3, there are 8 distinct combinations (subsets), including the empty set:

$$|\{\{\};\{1\};\{2\};\{1,2\};\{3\};\{1,3\};\{2,3\};\{1,2,3\}\}| = 2^3 = 8$$

Representing these subsets (in the same order) as base 2 numerals:

0. – 000
1. – 001
2. – 010
3. – 011
4. – 100
5. – 101
6. – 110
7. – 111

### Probability: Sampling a Random Combination

There are various algorithms to pick out a random combination from a given set or list. Rejection sampling is extremely slow for large sample sizes. One way to select a *k*-combination efficiently from a population of size *n* is to iterate across each element of the population, and at each step pick that element with a dynamically changing probability of $\frac{k - \#\text{samples chosen}}{n - \#\text{samples visited}}$.

## Pigeonhole Principle

In mathematics, the pigeonhole principle states that if $n$ items are put into $m$ containers, with $n > m$, then at least one container must contain more than one item. This theorem is exemplified in real life by truisms like "in any group of three gloves there must be at least two left gloves or at least two right gloves". It is an example of a counting argument. This seemingly obvious statement can be used to demonstrate possibly unexpected results; for example, that there must be (at least) two people in London who have the same number of hairs on their heads.

The first formalization of the idea is believed to have been made by Peter Gustav Lejeune Dirichlet in 1834 under the name *Schubfachprinzip* ("drawer principle" or "shelf principle"). For this reason it is also commonly called Dirichlet's box principle or Dirichlet's drawer principle.

Pigeons in holes. Here there are $n = 10$ pigeons in $m = 9$ holes. Since 10 is greater than 9, the pigeonhole principle says that at least one hole has more than one pigeon.

The principle has several generalizations and can be stated in various ways. In a more quantified version: for natural numbers $k$ and $m$, if $n = km + 1$ objects are distributed among $m$ sets, then the pigeonhole principle asserts that at least one of the sets will contain at least $k + 1$ objects. For arbitrary $n$ and $m$ this generalizes to $k+1 = \lfloor (n-1)/m \rfloor + 1 = \lceil n/m \rceil$, where $\lfloor \cdots \rfloor$ and $\lceil \cdots \rceil$ denote the floor and ceiling functions, respectively.

Though the most straightforward application is to finite sets (such as pigeons and boxes), it is also used with infinite sets that cannot be put into one-to-one correspondence. To do so requires the formal statement of the pigeonhole principle, which is *"there does not exist an injective function whose codomain is smaller than its domain"*. Advanced mathematical proofs like Siegel's lemma build upon this more general concept.

## Examples

### Sock-picking

Assume a drawer contains a mixture of black socks and blue socks, each of which can be worn on either foot, and that you are pulling a number of socks from the drawer without looking. What is the minimum number of pulled socks required to guarantee a pair of the same color? Using the pigeonhole principle, to have at least one pair of the same color ($m$ = 2 holes, one per color) using one pigeonhole per color, you need to pull only three socks from the drawer ($n$ = 3 items). Either you have *three* of one color, or you have *two* of one color and *one* of the other.

### Hand-shaking

If there are $n$ people who can shake hands with one another (where $n > 1$), the pigeonhole principle shows that there is always a pair of people who will shake hands with the same number of people. In this application of the principle, the 'hole' to which a person is assigned is the number of hands shaken by that person. Since each person shakes hands with some number of people from 0 to $n - 1$, there are $n$ possible holes. On the other hand, either the '0' hole or the '$n - 1$' hole or both must be empty, for it is impossible (if $n > 1$!) for some person to shake hands with everybody else while some person shakes hands with nobody. This leaves $n$ people to be placed into at most $n - 1$ non-empty holes, so that the principle applies.

### Hair-counting

We can demonstrate there must be at least two people in London with the same number of hairs on their heads as follows. Since a typical human head has an average of around 150,000 hairs, it is reasonable to assume (as an upper bound) that no one has more than 1,000,000 hairs on their head ($m$ = 1 million holes). There are more than 1,000,000 people in London ($n$ is bigger than 1 million items). Assigning a pigeonhole to each number of hairs on a person's head, and assign people to pigeonholes according to the number of hairs on their head, there must be at least two people assigned to the same pigeonhole by the 1,000,001st assignment (because they have the same number of hairs on their heads) (or, $n > m$). For the average case ($m$ = 150,000) with the constraint: fewest overlaps, there will be at most one person assigned to every pigeonhole and the 150,001st person assigned to the same pigeonhole as someone else. In the absence of this constraint, there may be empty pigeonholes because the "collision" happens before we get to the 150,001st person. The principle just proves the existence of an overlap; it says nothing of the number of overlaps (which falls under the subject of probability distribution).

## Birthday Problem

The birthday problem asks, for a set of $n$ randomly chosen people, what is the probability that some pair of them will have the same birthday? By the pigeonhole principle, if there are 367 people in the room, we know that there is at least one pair who share the same birthday, as there are only 366 possible birthdays to choose from (including February 29, if present). The birthday "paradox" refers to the result that even if the group is as small as 23 individuals, the probability that there is a pair of people with the same birthday is still above 50%. While at first glance this may seem surprising, it intuitively makes sense when considering that a comparison will actually be made between every possible pair of people rather than fixing one individual and comparing them solely to the rest of the group.

## Team Tournament

Imagine seven people who want to play in a tournament of teams ($n = 7$ items), with a limitation of only four teams ($m = 4$ holes) to choose from. The pigeonhole principle tells us that they cannot all play for different teams; there must be at least one team featuring at least two of the seven players:

$$\left\lfloor \frac{n-1}{m} \right\rfloor + 1 = \left\lfloor \frac{7-1}{4} \right\rfloor + 1 = \left\lfloor \frac{6}{4} \right\rfloor + 1 = 1 + 1 = 2$$

## Subset Sum

Any subset of size six from the set $S = \{1,2,3,...,9\}$ must contain two elements whose sum is 10. The pigeonholes will be labelled by the two element subsets $\{1,9\}$, $\{2,8\}$, $\{3,7\}$, $\{4,6\}$ and the singleton $\{5\}$, five pigeonholes in all. When the six "pigeons" (elements of the size six subset) are placed into these pigeonholes, each pigeon going into the pigeonhole that has it contained in its label, at least one of the pigeonholes labelled with a two element subset will have two pigeons in it.

## Uses and Applications

The principle can be used to prove that any lossless compression algorithm, provided it makes some inputs smaller (as the name compression suggests), will also make some other inputs larger. Otherwise, the set of all input sequences up to a given length $L$ could be mapped to the (much) smaller set of all sequences of length less than $L$ without collisions (because the compression is lossless), a possibility which the pigeonhole principle excludes.

A notable problem in mathematical analysis is, for a fixed irrational number $a$, to show that the set $\{[na]: n \text{ is an integer}\}$ of fractional parts is dense in $[0, 1]$. One finds that it is not easy to explicitly find integers $n$, $m$ such that $|na - ma| < e$, where $e > 0$ is a small positive number and $a$ is some arbitrary irrational number. But if one takes $M$ such that $1/M < e$, by the pigeonhole principle there must be $n_1, n_2 \in \{1, 2, ..., M + 1\}$ such that $n_1a$ and $n_2a$ are in the same integer subdivision of size $1/M$ (there are only $M$ such subdivisions between consecutive integers). In particular, we can find $n_1, n_2$ such that $n_1a$ is in $(p + k/M, p + (k + 1)/M)$, and $n_2a$ is in $(q + k/M, q + (k + 1)/M)$, for some $p, q$ integers and $k$ in $\{0, 1, ..., M - 1\}$. We can then easily verify that $(n_2 - n_1)a$ is in $(q - p - 1/M, q - p + 1/M)$. This implies that $[na] < 1/M < e$, where $n = n_2 - n_1$ or $n = n_1 - n_2$. This shows that 0 is a limit point of $\{[na]\}$. We can then use this fact to prove the case for $p$ in $(0, 1]$: find $n$ such that $[na] < 1/M < e$; then if $p \in (0, 1/M]$, we are done. Otherwise $p \in (j/M, (j + 1)/M]$, and by setting $k = \sup\{r \in N : r[na] < j/M\}$, one obtains $|[(k + 1)na] - p| < 1/M < e$.

Variants occurring in well known proofs: In the proof of the pumping lemma for regular languages, a version that mixes finite and infinite sets is used: If infinitely many objects are placed into finitely many boxes, then there exist two objects that share a box. In Fisk's solution of the Art gallery problem a sort of converse is used. If $n$ objects are placed into $k$ boxes, then there is a box containing at most $n/k$ objects.

## Alternate Formulations

The following are alternate formulations of the pigeonhole principle.

1. If $n$ objects are distributed over $m$ places, and if $n > m$, then some place receives at least two objects.
2. If $n$ objects are distributed over $n$ places in such a way that no place receives more than one object, then each place receives exactly one object.
3. If $n$ objects are distributed over $m$ places, and if $n < m$, then some place receives no object.
4. If $n$ objects are distributed over $n$ places in such a way that no place receives no object, then each place receives exactly one object.

## Strong Form

Let $q_1, q_2, ..., q_n$ be positive integers. If,

$$q_1 + q_2 + \cdots + q_n - n + 1$$

objects are distributed into $n$ boxes, then either the first box contains at least $q_1$ objects, or the second box contains at least $q_2$ objects, ..., or the $n$th box contains at least $q_n$ objects.

The simple form is obtained from this by taking $q_1 = q_2 = ... = q_n = 2$, which gives $n + 1$ objects. Taking $q_1 = q_2 = ... = q_n = r$ gives the more quantified version of the principle, namely:

Let $n$ and $r$ be positive integers. If $n(r - 1) + 1$ objects are distributed into $n$ boxes, then at least one of the boxes contains $r$ or more of the objects.

This can also be stated as, if $k$ discrete objects are to be allocated to $n$ containers, then at least one container must hold at least $\lceil k/n \rceil$ objects, where $\lceil x \rceil$ is the ceiling function, denoting the smallest integer larger than or equal to $x$. Similarly, at least one container must hold no more than $\lfloor k/n \rfloor$ objects, where $\lfloor x \rfloor$ is the floor function, denoting the largest integer smaller than or equal to $x$.

## Generalizations of the Pigeonhole Principle

A probabilistic generalization of the pigeonhole principle states that if $n$ pigeons are randomly put into $m$ pigeonholes with uniform probability $1/m$, then at least one pigeonhole will hold more than one pigeon with probability,

$$1-\frac{(m)_n}{m^n},$$

where $(m)_n$ is the falling factorial $m(m - 1)(m - 2)...(m - n + 1)$. For $n = 0$ and for $n = 1$ (and $m > 0$), that probability is zero; in other words, if there is just one pigeon, there cannot be a conflict. For $n > m$ (more pigeons than pigeonholes) it is one, in which case it coincides with the ordinary pigeonhole principle. But even if the number of pigeons does not exceed the number of pigeonholes ($n \leq m$), due to the random nature of the assignment of pigeons to pigeonholes there is often a substantial chance that clashes will occur. For example, if 2 pigeons are randomly assigned to 4 pigeonholes, there is a 25% chance that at least one pigeonhole will hold more than one pigeon; for 5 pigeons and 10 holes, that probability is 69.76%; and for 10 pigeons and 20 holes it is about 93.45%. If the number of holes stays fixed, there is always a greater probability of a pair when you add more pigeons. This problem is treated at much greater length in the birthday paradox.

A further probabilistic generalization is that when a real-valued random variable $X$ has a finite mean $E(X)$, then the probability is nonzero that $X$ is greater than or equal to $E(X)$, and similarly the probability is nonzero that $X$ is less than or equal to $E(X)$. To see that this implies the standard pigeonhole principle, take any fixed arrangement of $n$ pigeons into $m$ holes and let $X$ be the number of pigeons in a hole chosen uniformly at random. The mean of $X$ is $n/m$, so if there are more pigeons than holes the mean is greater than one. Therefore, $X$ is sometimes at least 2.

## Infinite Sets

The pigeonhole principle can be extended to infinite sets by phrasing it in terms of cardinal numbers: if the cardinality of set $A$ is greater than the cardinality of set $B$, then there is no injection from $A$ to $B$. However, in this form the principle is tautological, since the meaning of the statement that the cardinality of set $A$ is greater than the cardinality of set $B$ is exactly that there is no injective map from $A$ to $B$. However, adding at least one element to a finite set is sufficient to ensure that the cardinality increases.

Another way to phrase the pigeonhole principle for finite sets is similar to the principle that finite sets are Dedekind finite. Let $A$ and $B$ be finite sets. If there is a surjection from $A$ to $B$ that is not injective, then no surjection from $A$ to $B$ is injective. In fact no function of any kind from $A$ to $B$ is injective. This is not true for infinite sets. Consider the function on the natural numbers that sends 1 and 2 to 1, 3 and 4 to 2, 5 and 6 to 3, and so on.

There is a similar principle for infinite sets, if uncountably many pigeons are stuffed into countably many pigeonholes, there will exist at least one pigeonhole having uncountably many pigeons stuffed into it.

This principle is not a generalization of the pigeonhole principle for finite sets however, it is in general false for finite sets. In technical terms it says that if $A$ and $B$ are finite sets such that any surjective function from $A$ to $B$ is not injective, then there exists an element of $b$ of $B$ such that there exists a bijection between the preimage of $b$ and $A$. This is a quite different statement, and is absurd for large finite cardinalities.

## Inclusion–Exclusion Principle

In combinatorics, the inclusion–exclusion principle (also known as the sieve principle) is an equation relating the sizes of two sets and their union. It states that if A and B are two (finite) sets, then:

$$|A \cup B| = |A| + |B| - |A \cap B|.$$

The meaning of the statement is that the number of elements in the union of the two sets is the sum of the elements in each set, respectively, minus the number of elements that are in both. Similarly, for three sets A, B and C,

$$|A \cup B \cup C| = |A| + |B| + |C| - |A \cap B| - |A \cap C| - |B \cap C| + |A \cap B \cap C|.$$

This can be seen by counting how many times each region in the figure to the right is included in the right hand side. More generally, for finite sets $A_1$ , ..., $A_n$ , one has the identity:

$$\left|\bigcup_{i=1}^{n} A_i\right| = \sum_{i=1}^{n} |A_i| - \sum_{i,j:1 \le i \le j \le n} |A_i \cap A_j| + \sum_{i,j,k:1 \le i \le j \le k \le n} |A_i \cap A_j \cap A_k| - \ldots + (-1)^{n-1} |A_1 \cap \ldots \cap A_n|.$$

This can be compactly written as:

$$\left|\bigcup_{i=1}^{n} A_i\right| = \sum_{k=1}^{n} (-1)^{k+1} \left( \sum_{1 \le i_1 < \cdots < i_k \le n} |A_{i_1} \cap \ldots \cap A_{i_k}| \right)$$

The name comes from the idea that the principle is based on over-generous inclusion, followed by compensating exclusion. When n > 2 the exclusion of the pairwise intersections is (possibly) too severe, and the correct formula is as shown with alternating signs. This formula is attributed to Abraham de Moivre; it is sometimes also named for Daniel da Silva, Joseph Sylvester or Henri Poincaré.

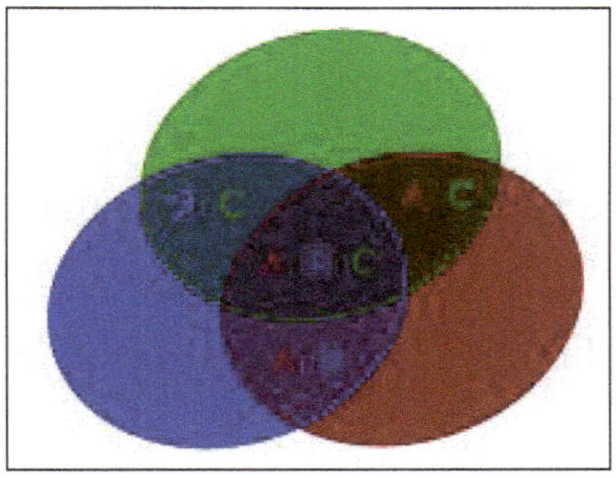

Inclusion–exclusion illustrated for three sets.

For the case of three sets A, B, C the inclusion–exclusion principle is illustrated in the graphic on the right.

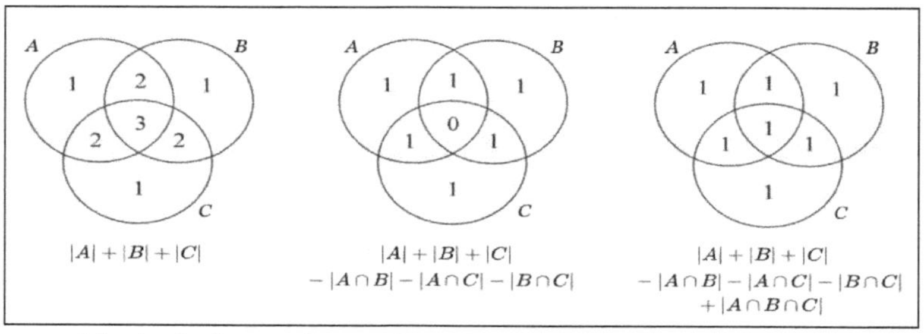


Each term of the inclusion-exclusion formula gradually corrects the count until finally each portion of the Venn Diagram is counted exactly once.

Let, A denote the union $\bigcup_i^n = 1 A_i$ of the sets $A_1, \ldots, A_n$. To prove the inclusion–exclusion principle in general, we first have to verify the identity.

$$1_A = \sum_{k=1}^{n} (-1)^{k+1} \sum_{\substack{I \subset \{1,\ldots,n\} \\ |I|=k}} 1_{A_I} \; (*)$$

for indicator functions, where,

$$A_I = \bigcap_{i \in I} A_i.$$

There are at least two ways to do this:

First possibility: It suffices to do this for every x in the union of $A_1, \ldots, A_n$. Suppose x belongs to exactly m sets with $1 \leq m \leq n$, for simplicity of notation say $A_1, \ldots, A_m$. Then the identity at x reduces to,

$$1 = \sum_{k=1}^{m} (-1)^{k+1} \sum_{\substack{I \subset \{1,\ldots,m\} \\ |I|=k}} 1.$$

The number of subsets of cardinality k of an m-element set is the combinatorical interpretation of the binomial coefficient $\binom{m}{k}$. Since $1 = \binom{m}{0}$, we have :

$$\binom{m}{0} = \sum_{k=1}^{m} (-1)^{k+1} \binom{m}{k}.$$

Putting all terms to the left-hand side of the equation, we obtain the expansion for $(1 - 1)^m$ given by the binomial theorem, hence we see that (*) is true for x.

Second possibility: The following function is identically zero:

$$\left(1_A - 1_{A_1}\right)\left(1_A - 1_{A_2}\right)\cdots\left(1_A - 1_{A_n}\right) = 0,$$

because: if x is not in A, then all factors are 0 - 0 = 0; and otherwise, if x does belong to some $A_m$, then the corresponding mth factor is 1 − 1 = 0. By expanding the product on the left-hand side, equation (*) follows.

Use of (*): To prove the inclusion–exclusion principle for the cardinality of sets, sum the equation (*) over all x in the union of $A_1$, ..., $A_n$. To derive the version used in probability, take the expectation in (*). In general, integrate the equation (*) with respect to μ. Always use linearity.

Alternative Proof: Pick an element contained in the union of all sets and let $A_1, A_2, \ldots, A_t$ be the individual sets containing it. (Note that $t > 0$.) Since the element is counted precisely once by the left-hand side of the equation, we need to show that it is counted only once by the right-hand side. By the binomial theorem,

$$(1-1)^t = \binom{t}{0} - \binom{t}{1} + \binom{t}{2} - \cdots + (-1)^t \binom{t}{t}$$

Using the convention that $\binom{t}{0} = 1$ and rearranging terms, we have:

$$\begin{aligned} 1 &= \binom{t}{1} - \binom{t}{2} + \cdots + (-1)^{t+1} \binom{t}{t} \\ &= \left|\{A_i \mid 1 \le i \le t\}\right| - \left|\{A_i \cap A_j \mid 1 \le i \le j \le t\}\right| + \cdots + (-1)^{t+1} \left|A_1 \cap A_2 \cap \cdots A_t\right| \end{aligned}$$

and so the chosen element is indeed counted only once by the right-hand side of the proposed equation.

Example: Suppose there is a deck of n cards, each card is numbered from 1 to n. Suppose a card numbered m is in the correct position if it is the mth card in the deck. How many ways, W, can the cards be shuffled with at least 1 card being in the correct position?

Begin by defining set $A_m$, which is all of the orderings of cards with the mth card correct. Then the number of orders, W, with at least one card being in the correct position, m, is:

$$W = \left|\bigcup_{m=1}^{n} A_m\right|.$$

Apply the principle of inclusion-exclusion,

$$\begin{aligned} W = &\sum_{m_1=1}^{n} \left|A_{m_1}\right| \\ &- \sum_{m_1, m_2 : 1 \le m_1 < m_2 \le n} \left|A_{m_1} \cap A_{m_2}\right| \\ &+ \sum_{m_1, m_2, m_3 : 1 \le m_1 < m_2 < m_3 \le n} \left|A_{m_1} \cap A_{m_2} \cap A_{m_3}\right| \\ &- \cdots \\ &+ (-1)^{p-1} \sum_{m_1, \ldots, m_p : 1 \le m_1 < \cdots m_p \le n} \left|A_{m_1} \cap \cdots \cap A_{m_p}\right| \\ &\cdots. \end{aligned}$$

Each value $A_{m_1} \cap \cdots \cap A_{m_p}$ represents the set of shuffles having p values $m_1$, ..., $m_p$ in the correct position. Note that the number of shuffles with p values correct only depends on p, not on the particular values of $m$. For example, the number of shuffles having the 1st, 3rd, and 17th cards in the correct position is the same as the number of shuffles having the 2nd, 5th, and 13th cards in the correct positions. It only matters that of the n cards, 3 were chosen to be in the correct position.

Thus there are $\binom{n}{p}$ terms in each summation.

$$
\begin{aligned}
W = & \binom{n}{1}|A_1| \\
& -\binom{n}{2}|A_1 \cap A_2| \\
& +\binom{n}{3}|A_1 \cap A_2 \cap A_3| \\
& -\cdots \\
& +(-1)^{p-1}\binom{n}{p}|A_1 \cap \cdots \cap A_p| \\
& \cdots.
\end{aligned}
$$

$|A_1 \cap \cdots \cap A_p|$ is the number of orderings having p elements in the correct position, which is equal to the number of ways of ordering the remaining n – p elements, or (n – p)!. Thus we finally get:

$$
\begin{aligned}
W = & \binom{n}{1}(n-1)! \\
& -\binom{n}{2}(n-2)! \\
& +\binom{n}{3}(n-3)! \\
& -\cdots \\
& +(-1)^{p-1}\binom{n}{p}(n-p)! \\
& \cdots
\end{aligned}
$$

$$W = \sum_{p=1}^{n}(-1)^{p-1}\binom{n}{p}(n-p)!.$$

Noting that $\binom{n}{p} = \frac{n!}{p!(n-p)!}$.

This Reduces to:

$$W = \sum_{p=1}^{n} (-1)^{p-1} \frac{n!}{p!}.$$

A permutation where no card is in the correct position is called a derangement. Taking n! to be the total number of permutations, the probability Q that a random shuffle produces a derangement is given by,

$$Q = 1 - \frac{W}{n!} = \sum_{p=0}^{n} \frac{(-1)^p}{p!},$$

the Taylor expansion of $e^{-1}$. Thus the probability of guessing an order for a shuffled deck of cards and being incorrect about every card is approximately 1/e or 37%.

## References


- High School Textbook for full-time student (Required) Mathematics Book II B (in Chinese) (2nd ed.). China: People's Education Press. June 2006. Pp. 107–116. ISBN 978-7-107-19616-4
- Combinatorics, science: britannica.com, Retrieved 19 April, 2019
- Knuth, D. E. (2005), "Generating All Combinations and Partitions", The Art of Computer Programming, 4, Fascicle 3, Addison-Wesley, pp. 5–6, ISBN 0-201-85394-9
- Arndt, Jörg (2011). Matters Computational. Ideas, Algorithms, Source Code. Springer. Doi:10.1007/978-3-642-14764-7
- Gross, Jonathan L. (2008), Combinatorial Methods with Computer Applications, Chapman & Hall/CRC, p. 29, ISBN 978-1-58488-743-0
- Broemeling, Lyle D. (1 November 2011). "An Account of Early Statistical Inference in Arab Cryptology". The American Statistician. 65 (4): 255–257. Doi:10.1198/tas.2011.10191

# 5
# Graph Theory

The relation between the pairwise objects can be modeled and studied under graph theory. Graph factorization, graph isomorphism, graph automorphism, graph amalgamation, etc. are some of the concepts of graph theory. All these concepts of graph theory and types of graphs have been carefully analyzed in this chapter.

A graph is a structure amounting to a set of objects in which some pairs of the objects are in some sense "related". The objects of the graph correspond to vertices and the relations between them correspond to edges. A graph is depicted diagrammatically as a set of dots depicting vertices connected by lines or curves depicting edges. Formally,

> "A graph $G=(V,E)$ consists of $V$, a non-empty set of vertices (or nodes) and $E$, a set of edges. Each edge has either one or two vertices associated with it, called its endpoints."

Graph theory is branch of mathematics concerned with networks of points connected by lines. The subject of graph theory had its beginnings in recreational math problems, but it has grown into a significant area of mathematical research, with applications in chemistry, operations research, social sciences, and computer science.

The history of graph theory may be specifically traced to 1735, when the Swiss mathematician Leonhard Euler solved the Königsberg bridge problem. The Königsberg bridge problem was an old puzzle concerning the possibility of finding a path over every one of seven bridges that span a forked river flowing past an island—but without crossing any bridge twice. Euler argued that no such path exists. His proof involved only references to the physical arrangement of the bridges, but essentially he proved the first theorem in graph theory.

As used in graph theory, the term graph does not refer to data charts, such as line graphs or bar graphs. Instead, it refers to a set of vertices (that is, points or nodes) and of edges (or lines) that connect the vertices. When any two vertices are joined by more than one edge, the graph is called a multigraph. A graph without loops and with at most one edge between any two vertices is called a simple graph. Unless stated otherwise, graph is assumed to refer to a simple graph. When each vertex is connected by an edge to every other vertex, the graph is called a complete graph. When appropriate, a direction may be assigned to each edge to produce what is known as a directed graph, or digraph.

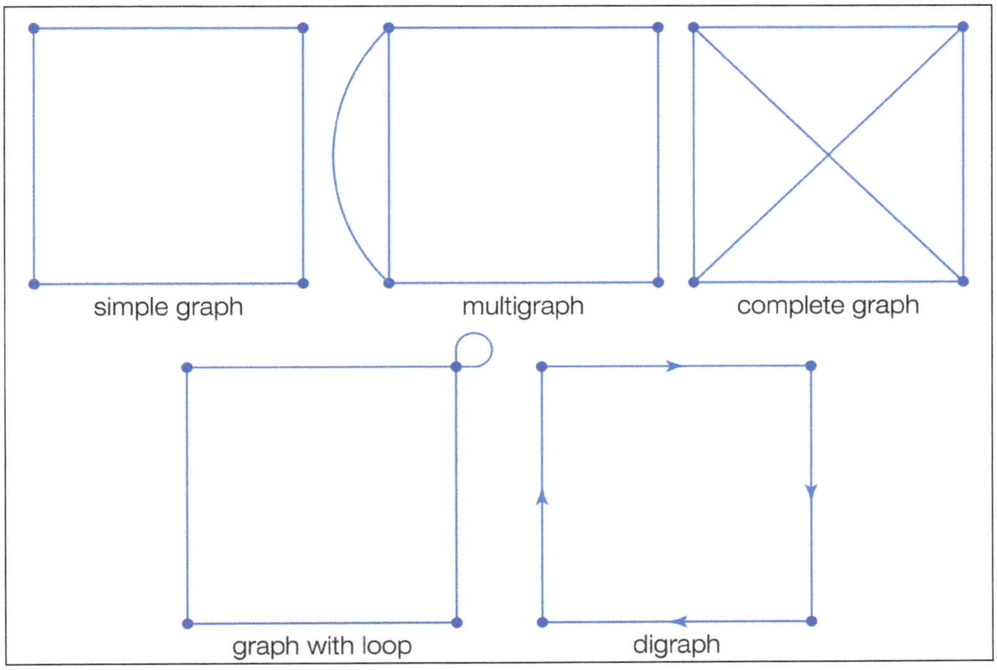


Basic types of graphs.

An important number associated with each vertex is its degree, which is defined as the number of edges that enter or exit from it. Thus, a loop contributes 2 to the degree of its vertex. For instance, the vertices of the simple graph shown in the diagram all have a degree of 2, whereas the vertices of the complete graph shown are all of degree 3. Knowing the number of vertices in a complete graph characterizes its essential nature. For this reason, complete graphs are commonly designated *Kn*, where *n* refers to the number of vertices, and all vertices of *Kn* have degree $n - 1$. Translated into the terminology of modern graph theory, Euler's theorem about the Königsberg bridge problem could be restated as follows: If there is a path along edges of a multigraph that traverses each edge once and only once, then there exist at most two vertices of odd degree; furthermore, if the path begins and ends at the same vertex, then no vertices will have odd degree.

Another important concept in graph theory is the path, which is any route along the edges of a graph. A path may follow a single edge directly between two vertices, or it may follow multiple edges through multiple vertices. If there is a path linking any two vertices in a graph, that graph is said to be connected. A path that begins and ends at the same vertex without traversing any edge more than once is called a circuit, or a closed path. A circuit that follows each edge exactly once while visiting every vertex is known as an Eulerian circuit, and the graph is called an Eulerian graph. An Eulerian graph is connected and, in addition, all its vertices have even degree.

A graph is a collection of vertices, or nodes, and edges between some or all of the vertices. When there exists a path that traverses each edge exactly once such that the path begins and ends at the same vertex, the path is known as an Eulerian circuit, and the graph is known as an Eulerian graph. *Eulerian* refers to the Swiss mathematician Leonhard Euler, who invented graph theory in the 18th century.

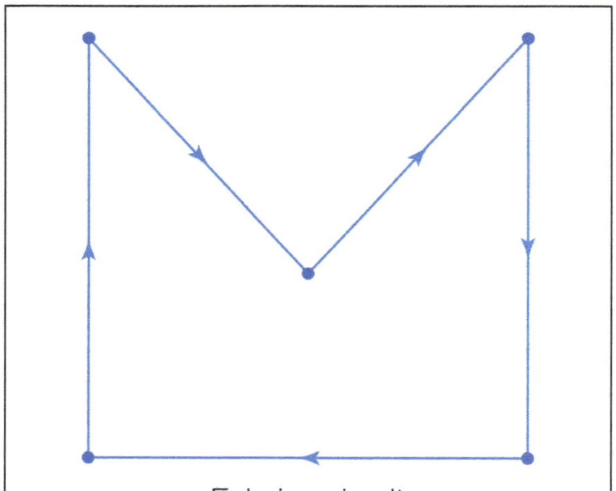

Eulerian circuit

In 1857 the Irish mathematician William Rowan Hamilton invented a puzzle (the Icosian Game) that he later sold to a game manufacturer for £25. The puzzle involved finding a special type of path, later known as a Hamiltonian circuit, along the edges of a dodecahedron (a Platonic solid consisting of 12 pentagonal faces) that begins and ends at the same corner while passing through each corner exactly once. The knight's tour is another example of a recreational problem involving a Hamiltonian circuit. Hamiltonian graphs have been more challenging to characterize than Eulerian graphs, since the necessary and sufficient conditions for the existence of a Hamiltonian circuit in a connected graph are still unknown.

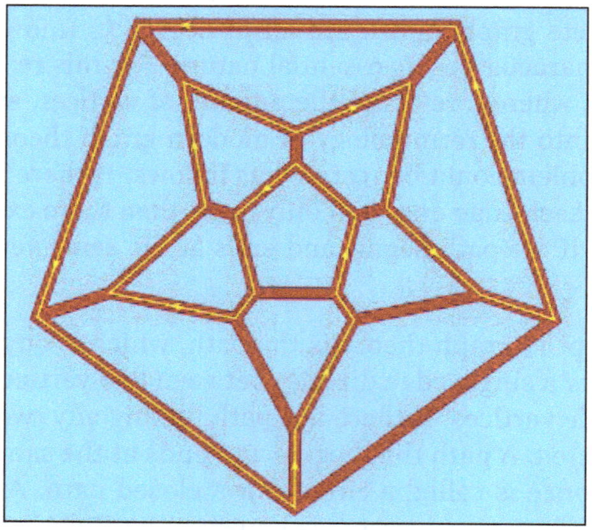

Hamiltonian circuit

A directed graph in which the path begins and ends on the same vertex (a closed loop) such that each vertex is visited exactly once is known as a Hamiltonian circuit. The 19th-century Irish mathematician William Rowan Hamilton began the systematic mathematical study of such graphs.

The histories of graph theory and topology are closely related, and the two areas share many common problems and techniques. Euler referred to his work on the Königsberg bridge problem as an example of *geometria situs*—the "geometry of position"—while the development of topological ideas during the second half of the 19th century became known as *analysis situs*—the "analysis of position." In 1750 Euler discovered the polyhedral formula $V - E + F = 2$ relating the number of vertices ($V$), edges ($E$), and faces ($F$) of a polyhedron (a solid, whose faces are polygons). The vertices and edges of a polyhedron form a graph on its surface, and this notion led to consideration of graphs on other surfaces such as a torus (the surface of a solid doughnut) and how they divide the surface into disklike faces. Euler's formula was soon generalized to surfaces as $V - E + F = 2 - 2g$, where $g$ denotes the genus, or number of "doughnut holes," of the surface. Having considered a surface divided into polygons by an embedded graph, mathematicians began to study ways of constructing surfaces, and later more general spaces, by pasting polygons together. This was the beginning of the field of combinatorial topology, which later, through the work of the French mathematician Henri Poincaré and others, grew into what is known as algebraic topology.

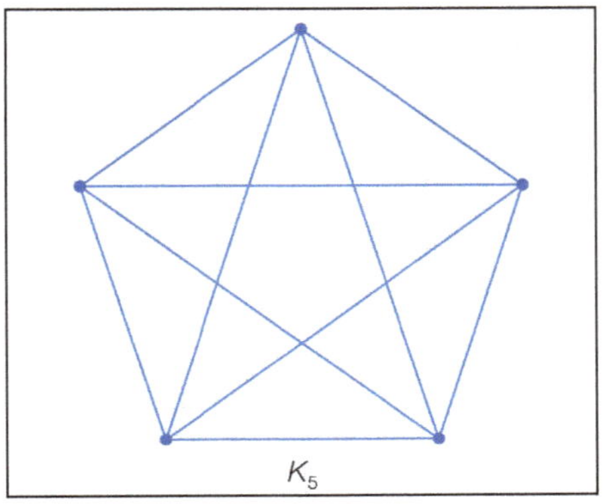


$K_5$ is not a planar graph, because there does not exist any way to connect every vertex to every other vertex with edges in the plane such that no edges intersect.

The connection between graph theory and topology led to a subfield called topological graph theory. An important problem in this area concerns planar graphs. These are graphs that can be drawn as dot-and-line diagrams on a plane (or, equivalently, on a sphere) without any edges crossing except at the vertices where they meet. Complete graphs with four or fewer vertices are planar, but complete graphs with five vertices ($K_5$) or more are not. Nonplanar graphs cannot be drawn on a plane or on the surface of a sphere without edges intersecting each other between the vertices. The use of diagrams of dots and lines to represent graphs actually grew out of 19th-century chemistry, where lettered vertices denoted individual atoms and connecting lines denoted chemical bonds (with degree corresponding to valence), in which planarity had important chemical consequences. The first use, in this context, of the word *graph* is attributed to the 19th-century Englishman James Sylvester, one of several mathematicians interested in counting special types of diagrams representing molecules.

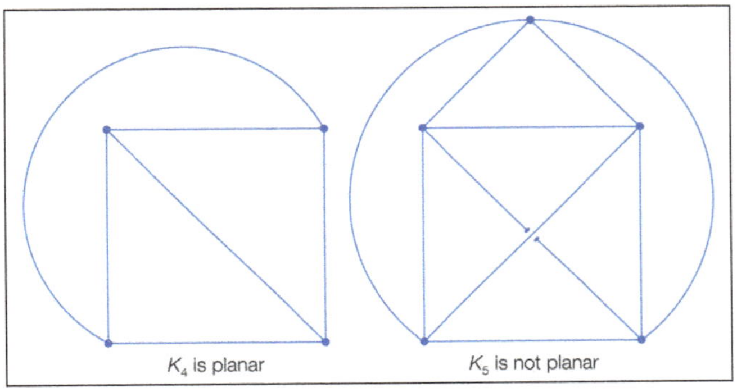


Planar Graph; Nonplanar Graph

With fewer than five vertices in a two-dimensional plane, a collection of paths between vertices can be drawn in the plane such that no paths intersect. With five or more vertices in a two-dimensional plane, a collection of nonintersecting paths between vertices cannot be drawn without the use of a third dimension.

Another class of graphs is the collection of the complete bipartite graphs $K_{m,n}$, which consist of the simple graphs that can be partitioned into two independent sets of $m$ and $n$ vertices such that there are no edges between vertices within each set and every vertex in one set is connected by an edge to every vertex in the other set. Like $K_5$, the bipartite graph $K_{3,3}$ is not planar, disproving a claim made in 1913 by the English recreational problemist Henry Dudeney to a solution to the "gas-water-electricity" problem. In 1930 the Polish mathematician Kazimierz Kuratowski proved that any nonplanar graph must contain a certain type of copy of $K_5$ or $K_{3,3}$. While $K_5$ and $K_{3,3}$ cannot be embedded in a sphere, they can be embedded in a torus. The graph-embedding problem concerns the determination of surfaces in which a graph can be embedded and thereby generalizes the planarity problem. It was not until the late 1960s that the embedding problem for the complete graphs *Kn* was solved for all *n*.

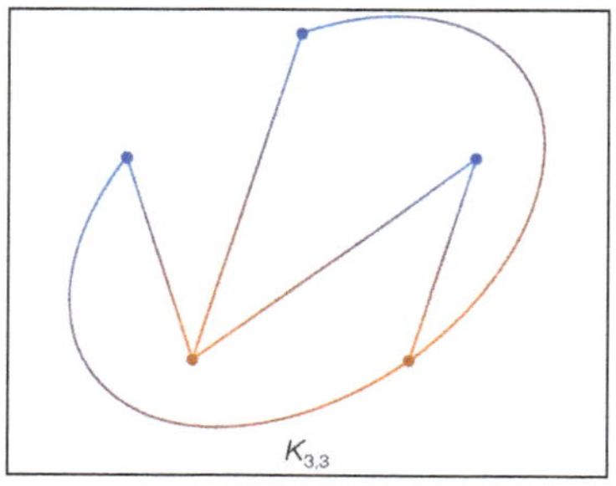


A bipartite map, such as $K_{3,3}$, consists of two sets of points in the two-dimensional plane such that every vertex in one set (the set of red vertices) can be connected to every vertex in the other set (the set of blue vertices) without any of the paths intersecting.

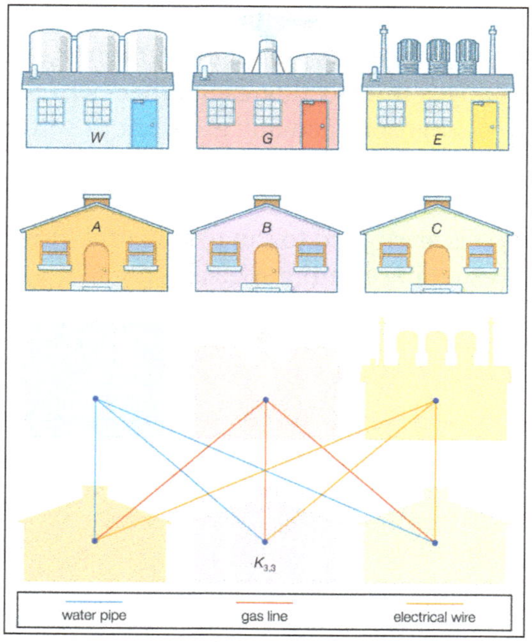


Dudeney puzzle.

The English recreational problemist Henry Dudeney claimed to have a solution to a problem that he posed in 1913 that required each of three houses to be connected to three separate utilities such that no utility service pipes intersected. Dudeney's solution involved running a pipe through one of the houses, which would not be considered a valid solution in graph theory. In a two-dimensional plane, a collection of six vertices (shown here as the vertices in the homes and utilities) that can be split into two completely separate sets of three vertices (that is, the vertices in the three homes and the vertices in the three utilities) is designated a $K_{3,3}$ bipartite graph. The two parts of such graphs cannot be interconnected within the two-dimensional plane without intersecting some paths.

Another problem of topological graph theory is the map-colouring problem. This problem is an outgrowth of the well-known four-colour map problem, which asks whether the countries on every map can be coloured by using just four colours in such a way that countries sharing an edge have different colours. Asked originally in the 1850s by Francis Guthrie, then a student at University College London, this problem has a rich history filled with incorrect attempts at its solution. In an equivalent graph-theoretic form, one may translate this problem to ask whether the vertices of a planar graph can always be coloured by using just four colours in such a way that vertices joined by an edge have different colours. The result was finally proved in 1976 by using computerized checking of nearly 2,000 special configurations. Interestingly, the corresponding colouring problem concerning the number of colours required to colour maps on surfaces of higher genus was completely solved a few years earlier; for example, maps on a torus may require as many as seven colours. This work confirmed that a formula of the English mathematician Percy Heawood from 1890 correctly gives these colouring numbers for all surfaces except the one-sided surface known as the Klein bottle, for which the correct colouring number had been determined in 1934.

Among the current interests in graph theory are problems concerning efficient algorithms for finding optimal paths (depending on different criteria) in graphs. Two well-known examples are the Chinese postman problem (the shortest path that visits each edge at least once), which was solved in the 1960s, and the traveling salesman problem (the shortest path that begins and ends at the same vertex and visits each edge exactly once), which continues to attract the attention of many researchers because of its applications in routing data, products, and people. Work on such problems is related to the field of linear programming, which was founded in the mid-20th century by the American mathematician George Dantzig.

## Vertex

In mathematics, and more specifically in graph theory, a vertex (plural vertices) or node is the fundamental unit of which graphs are formed: an undirected graph consists of a set of vertices and a set of edges (unordered pairs of vertices), while a directed graph consists of a set of vertices and a set of arcs (ordered pairs of vertices). In a diagram of a graph, a vertex is usually represented by a circle with a label, and an edge is represented by a line or arrow extending from one vertex to another.

From the point of view of graph theory, vertices are treated as featureless and indivisible objects, although they may have additional structure depending on the application from which the graph arises; for instance, a semantic network is a graph in which the vertices represent concepts or classes of objects.

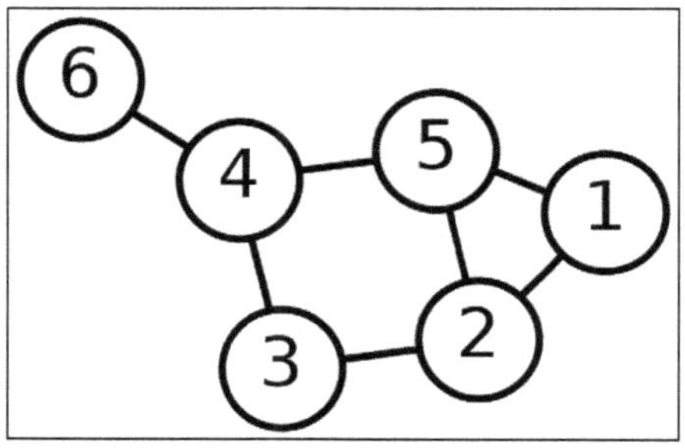


A graph with 6 vertices and 7 edges where the vertex number 6 on the far-left is a leaf vertex or a pendant vertex.

The two vertices forming an edge are said to be the endpoints of this edge, and the edge is said to be incident to the vertices. A vertex $w$ is said to be adjacent to another vertex $v$ if the graph contains an edge $(v,w)$. The neighborhood of a vertex $v$ is an induced subgraph of the graph, formed by all vertices adjacent to $v$.

## Types of Vertices

The degree of a vertex, denoted (v) in a graph is the number of edges incident to it. An isolated vertex is a vertex with degree zero; that is, a vertex that is not an endpoint of any edge (the example image illustrates one isolated vertex). A leaf vertex (also pendant vertex) is a vertex with degree one. In a directed graph, one can distinguish the outdegree (number of outgoing edges), denoted $\delta^+(v)$, from the indegree (number of incoming edges), denoted $\delta^-(v)$; a source vertex is a vertex with indegree zero, while a sink vertex is a vertex with outdegree zero. A simplicial vertex is one whose neighbors form a clique: every two neighbors are adjacent. A universal vertex is a vertex that is adjacent to every other vertex in the graph.

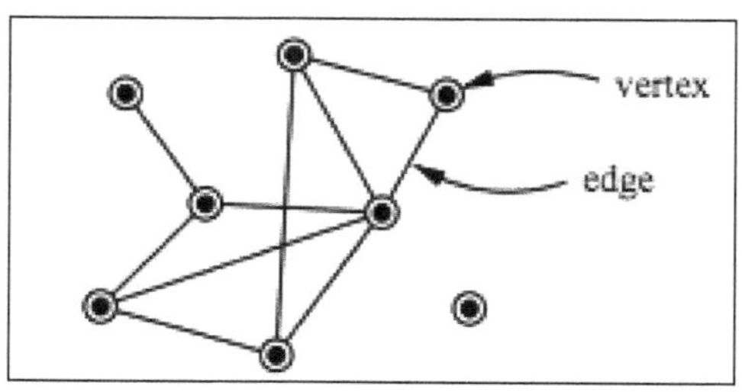


Example network with 8 vertices (of which one is isolated) and 10 edges.

A cut vertex is a vertex the removal of which would disconnect the remaining graph; a vertex separator is a collection of vertices the removal of which would disconnect the remaining graph into small pieces. A k-vertex-connected graph is a graph in which removing fewer than $k$ vertices always leaves the remaining graph connected. An independent set is a set of vertices no two of which are adjacent, and a vertex cover is a set of vertices that includes at least one endpoint of each edge in the graph. The vertex space of a graph is a vector space having a set of basis vectors corresponding with the graph's vertices.

A graph is vertex-transitive if it has symmetries that map any vertex to any other vertex. In the context of graph enumeration and graph isomorphism it is important to distinguish between labeled vertices and unlabeled vertices. A labeled vertex is a vertex that is associated with extra information that enables it to be distinguished from other labeled vertices; two graphs can be considered isomorphic only if the correspondence between their vertices pairs up vertices with equal labels. An unlabeled vertex is one that can be substituted for any other vertex based only on its adjacencies in the graph and not based on any additional information.

Vertices in graphs are analogous to, but not the same as, vertices of polyhedra: the skeleton of a polyhedron forms a graph, the vertices of which are the vertices of the polyhedron, but polyhedron vertices have additional structure (their geometric location) that is not assumed to be present in graph theory. The vertex figure of a vertex in a polyhedron is analogous to the neighborhood of a vertex in a graph.

## Degree of Vertex

In graph theory, the degree (or valency) of a vertex of a graph is the number of edges that are incident to the vertex, and in a multigraph, loops are counted twice. The degree of a vertex $v$ is denoted $\deg(v)$ or $\deg v$. The maximum degree of a graph $G$, denoted by $\Delta(G)$, and the minimum degree of a graph, denoted by $\delta(G)$, are the maximum and minimum degree of its vertices. In the multigraph on the right, the maximum degree is 5 and the minimum degree is 0.

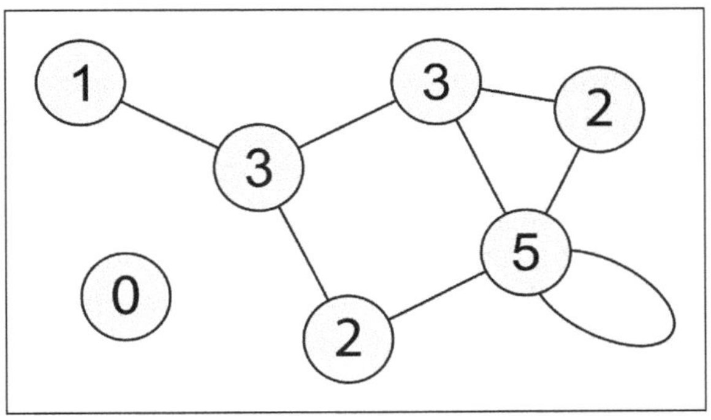

A multigraph with vertices labeled by degree.

In a regular graph, every vertex has the same degree, and so we can speak of *the* degree of the graph. A complete graph (denoted $K_n$, where $n$ is the number of vertices in the graph) is a special kind of regular graph where all vertices have the maximum degree, $n-1$.

## Handshaking Lemma

The degree sum formula states that, given a graph $G=(V,E)$,

$$\sum_{v\in V}\deg(v)=2|E|.$$

The formula implies that in any undirected graph, the number of vertices with odd degree is even. This statement (as well as the degree sum formula) is known as the handshaking lemma. The latter name comes from a popular mathematical problem, to prove that in any group of people the number of people who have shaken hands with an odd number of other people from the group is even.

## Degree Sequence

The degree sequence of an undirected graph is the non-increasing sequence of its vertex degrees; for the above graph it is (5, 3, 3, 2, 2, 1, 0). The degree sequence is a graph invariant so isomorphic graphs have the same degree sequence. However, the degree sequence does not, in general, uniquely identify a graph; in some cases, non-isomorphic graphs have the same degree sequence.

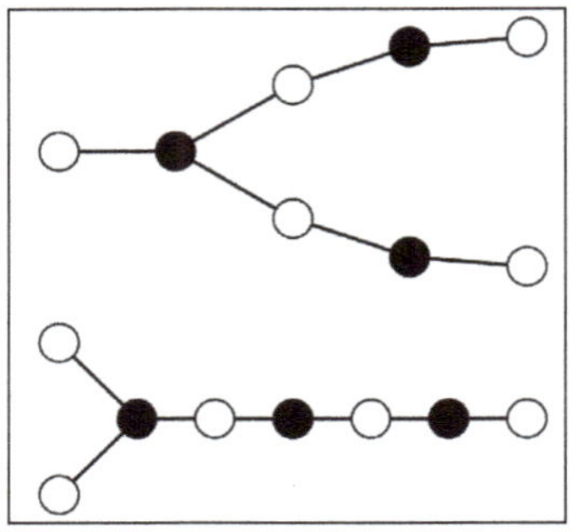

Two non-isomorphic graphs with the same degree sequence (3, 2, 2, 2, 2, 1, 1, 1).

The degree sequence problem is the problem of finding some or all graphs with the degree sequence being a given non-increasing sequence of positive integers. (Trailing zeroes may be ignored since they are trivially realized by adding an appropriate number of isolated vertices to the graph.) A sequence which is the degree sequence of some graph, i.e. for which the degree sequence problem has a solution, is called a graphic or graphical sequence. As a consequence of the degree sum formula, any sequence with an odd sum, such as (3, 3, 1), cannot be realized as the degree sequence of a graph. The converse is also true: if a sequence has an even sum, it is the degree sequence of a multigraph. The construction of such a graph is straightforward: connect vertices with odd degrees in pairs by a matching, and fill out the remaining even degree counts by self-loops. The question of whether a given degree sequence can be realized by a simple graph is more challenging. This problem is also called graph realization problem and can either be solved by the Erdős–Gallai theorem or the Havel–Hakimi algorithm. The problem of finding or estimating the number of graphs with a given degree sequence is a problem from the field of graph enumeration.

## Special Values

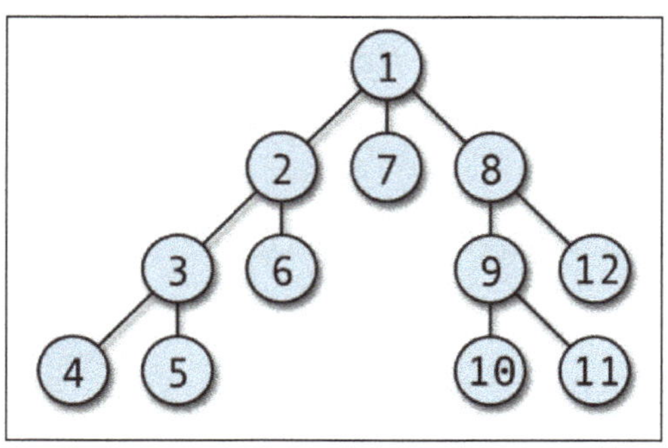


An undirected graph with leaf nodes 4, 5, 6, 7, 10, 11, and 12.

- A vertex with degree 0 is called an isolated vertex.
- A vertex with degree 1 is called a leaf vertex or end vertex, and the edge incident with that vertex is called a pendant edge. In the graph on the right, {3,5} is a pendant edge. This

terminology is common in the study of trees in graph theory and especially trees as data structures.

- A vertex with degree $n-1$ in a graph on $n$ vertices is called a dominating vertex.

## Global Properties

- If each vertex of the graph has the same degree $k$ the graph is called a $k$-regular graph and the graph itself is said to have degree $k$. Similarly, a bipartite graph in which every two vertices on the same side of the bipartition as each other have the same degree is called a biregular graph.
- An undirected, connected graph has an Eulerian path if and only if it has either 0 or 2 vertices of odd degree. If it has 0 vertices of odd degree, the Eulerian path is an Eulerian circuit.
- A directed graph is a pseudoforest if and only if every vertex has outdegree at most 1. A functional graph is a special case of a pseudoforest in which every vertex has outdegree exactly 1.
- By Brooks' theorem, any graph other than a clique or an odd cycle has chromatic number at most $\Delta$, and by Vizing's theorem any graph has chromatic index at most $\Delta + 1$.
- A $k$-degenerate graph is a graph in which each subgraph has a vertex of degree at most $k$.

# Vertex Separator

In graph theory, a vertex subset $S \subset V$ is a vertex separator (or vertex cut, separating set) for nonadjacent vertices $a$ and $b$ if the removal of $S$ from the graph separates $a$ and $b$ into distinct connected components.

## Examples

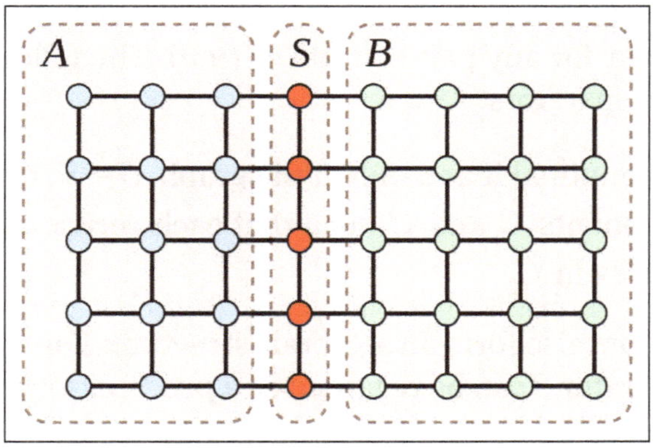


A separator for a grid graph.

Consider a grid graph with $r$ rows and $c$ columns; the total number $n$ of vertices is $r*c$. For instance, in the illustration, $r = 5$, $c = 8$, and $n = 40$. If $r$ is odd, there is a single central row, and otherwise there are two rows equally close to the center; similarly, if $c$ is odd, there is a single central column, and otherwise there are two columns equally close to the center. Choosing $S$ to be any of these central rows or columns, and removing $S$ from the graph, partitions the graph into two smaller connected subgraphs $A$ and $B$, each of which has at most $n/2$ vertices. If $r \leq c$ (as in the illustration), then choosing a central column will give a separator $S$ with $r \leq \sqrt{n}$ vertices, and similarly if $c \leq r$ then choosing a central row will give a separator with at most $\sqrt{n}$ vertices. Thus, every grid graph has a separator $S$ of size at most $\sqrt{n}$, the removal of which partitions it into two connected components, each of size at most $n/2$.

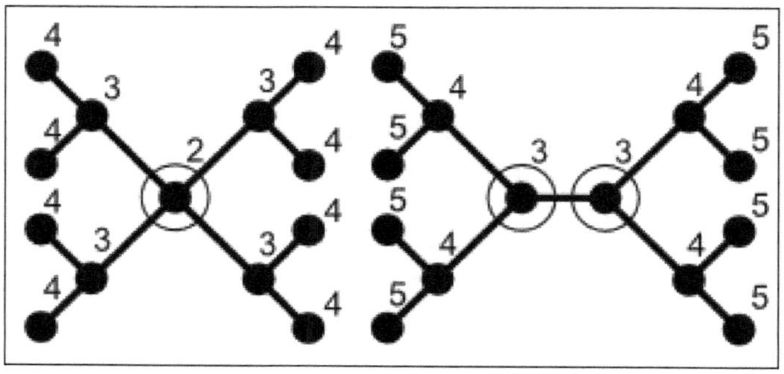


On the left a centered tree, on the right a bicentered one. The numbers show each node's eccentricity.

To give another class of examples, every free tree $T$ has a separator $S$ consisting of a single vertex, the removal of which partitions $T$ into two or more connected components, each of size at most $n/2$. More precisely, there is always exactly one or exactly two vertices, which amount to such a separator, depending on whether the tree is centered or bicentered.

As opposed to these examples, not all vertex separators are *balanced*, but that property is most useful for applications in computer science, such as the planar separator theorem.

## Minimal Separators

Let $S$ be an *(a,b)*-separator, that is, a vertex subset that separates two nonadjacent vertices $a$ and $b$. Then $S$ is a *minimal (a,b)-separator* if no proper subset of $S$ separates $a$ and $b$. More generally, $S$ is called a *minimal separator* if it is a minimal separator for some pair *(a,b)* of nonadjacent vertices. Notice that this is different from *minimal separating set* which says that no proper subset of $S$ is a minimal *(u,v)*-separator for any pair of vertices *(u,v)*. The following is a well-known result characterizing the minimal separators:

A vertex separator $S$ in $G$ is minimal if and only if the graph $G-S$, obtained by removing $S$ from $G$, has two connected components $C_1$ and $C_2$ such that each vertex in $S$ is both adjacent to some vertex in $C_1$ and to some vertex in $C_2$.

The minimal "(a,b)"-separators also form an algebraic structure: For two fixed vertices $a$ and $b$ of a given graph $G$, an *(a,b)*-separator $S$ can be regarded as a *predecessor* of another (a,b)-separator $T$, if every path from $a$ to $b$ meets $S$ before it meets $T$. More rigorously, the predecessor relation is defined as follows: Let $S$ and $T$ be two *(a,b)*-separators in 'G'. Then $S$ is a predecessor of $T$, in symbols $S \sqsubseteq_{a,b}^{G} T$, if for each $x \in S \setminus T$, every path connecting $x$ to $b$ meets $T$. It follows from the definition that the predecessor relation yields a preorder on the set of all *(a,b)*-separators. Furthermore,

Escalante (1972) proved that the predecessor relation gives rise to a complete lattice when restricted to the set of *minimal (a,b)*-separators in *G*.

## Vertex Cover

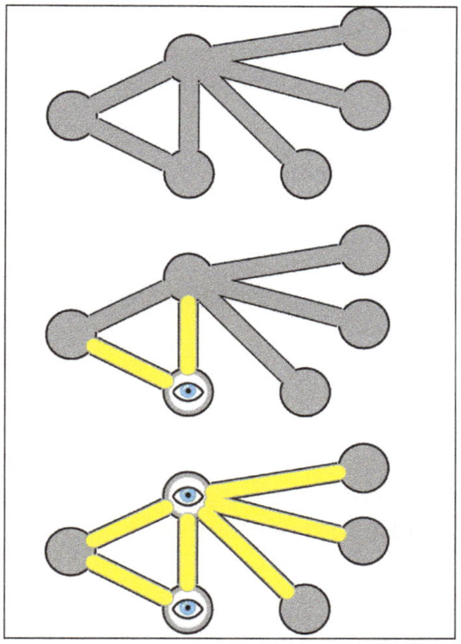

Example graph that has a vertex cover comprising 2 vertices (bottom), but none with fewer.

In the mathematical discipline of graph theory, a vertex cover (sometimes node cover) of a graph is a set of vertices such that each edge of the graph is incident to at least one vertex of the set. The problem of finding a minimum vertex cover is a classical optimization problem in computer science and is a typical example of an NP-hard optimization problem that has an approximation algorithm. Its decision version, the vertex cover problem, was one of Karp's 21 NP-complete problems and is therefore a classical NP-complete problem in computational complexity theory. Furthermore, the vertex cover problem is fixed-parameter tractable and a central problem in parameterized complexity theory.

The minimum vertex cover problem can be formulated as a half-integral linear program whose dual linear program is the maximum matching problem.

Formally, a vertex cover $V'$ of an undirected graph $G=(V,E)$ is a subset of $V$ such that $uv \in E \Rightarrow u \in V' \vee v \in V'$, that is to say it is a set of vertices $V'$ where every edge has at least one endpoint in the vertex cover $V'$. Such a set is said to *cover* the edges of $G$. The following figure shows two examples of vertex covers, with some vertex cover $V'$ marked in red.

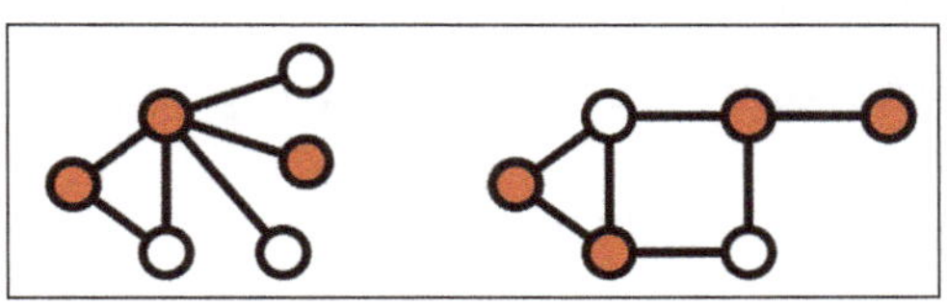

A *minimum vertex cover* is a vertex cover of smallest possible size. The vertex cover number $\tau$ is the size of a minimum vertex cover, i.e. $\tau = |V'|$. The following figure shows examples of minimum vertex covers in the previous graphs.

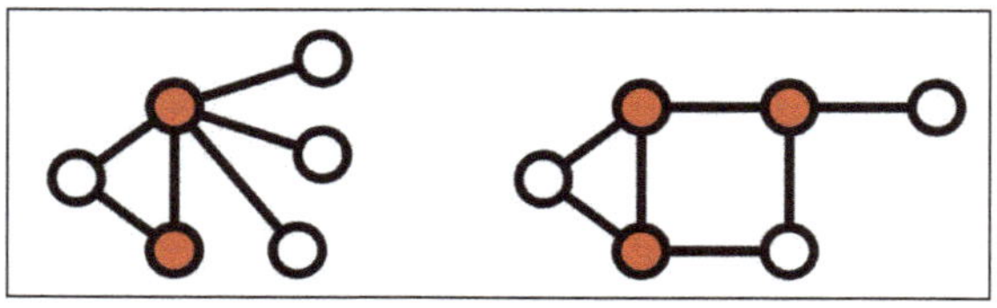

## Examples

- The set of all vertices is a vertex cover.
- The endpoints of any maximal matching form a vertex cover.
- The complete bipartite graph $K_{m,n}$ has a minimum vertex cover of size $\tau(K_{m,n}) = \min\{m,n\}$.

## Properties

- A set of vertices is a vertex cover if and only if its complement is an independent set.
- Consequently, the number of vertices of a graph is equal to its minimum vertex cover number plus the size of a maximum independent set.

## Computational Problem

The minimum vertex cover problem is the optimization problem of finding a smallest vertex cover in a given graph.

INSTANCE: Graph $G$

OUTPUT: Smallest number $k$ such that $G$ has a vertex cover of size $k$.

If the problem is stated as a decision problem, it is called the vertex cover problem:

INSTANCE: Graph $G$ and positive integer $k$.

QUESTION: Does $G$ have a vertex cover of size at most $k$?

The vertex cover problem is an NP-complete problem: it was one of Karp's 21 NP-complete problems. It is often used in computational complexity theory as a starting point for NP-hardness proofs.

## ILP Formulation

Assume that every vertex has an associated cost of $c(v) \geq 0$. The (weighted) minimum vertex cover problem can be formulated as the following integer linear program (ILP).

minimize $\sum_{v \in V} c(v) x_v$ (minimize the total cost) subject to $x_u + x_v \geq 1$ for all $\{u,v\} \in E$ (cover every edge of the graph)

$x_v \in \{0,1\}$ for all $v \in V$. (every vertex is either in the vertex cover or not)

This ILP belongs to the more general class of ILPs for covering problems. The integrality gap of this ILP is $2$, so its relaxation gives a factor-$2$ approximation algorithm for the minimum vertex cover problem. Furthermore, the linear programming relaxation of that ILP is *half-integral*, that is, there exists an optimal solution for which each entry $x_v$ is either 0, 1/2, or 1.

## Exact Evaluation

The decision variant of the vertex cover problem is NP-complete, which means it is unlikely that there is an efficient algorithm to solve it exactly for arbitrary graphs. NP-completeness can be proven by reduction from 3-satisfiability or, as Karp did, by reduction from the clique problem. Vertex cover remains NP-complete even in cubic graphs and even in planar graphs of degree at most 3.

For bipartite graphs, the equivalence between vertex cover and maximum matching described by Kőnig's theorem allows the bipartite vertex cover problem to be solved in polynomial time.

For tree graphs, an algorithm finds a minimal vertex cover in polynomial time by finding the first leaf in the tree and adding its parent to the minimal vertex cover, then deleting the leaf and parent and all associated edges and continuing repeatedly until no nodes remain in the tree.

## Fixed-parameter Tractability

An exhaustive search algorithm can solve the problem in time $2^k n^{O(1)}$, where $k$ is the size of the vertex cover. Vertex cover is therefore fixed-parameter tractable, and if we are only interested in small $k$, we can solve the problem in polynomial time. One algorithmic technique that works here is called *bounded search tree algorithm*, and its idea is to repeatedly choose some vertex and recursively branch, with two cases at each step: place either the current vertex or all its neighbours into the vertex cover. The algorithm for solving vertex cover that achieves the best asymptotic dependence on the parameter runs in time $O(1.2738^k + (k \cdot n))$. The klam value of this time bound (an estimate for the largest parameter value that could be solved in a reasonable amount of time) is approximately 190. That is, unless additional algorithmic improvements can be found, this algorithm is suitable only for instances whose vertex cover number is 190 or less. Under reasonable complexity-theoretic assumptions, namely the exponential time hypothesis, the running time cannot be improved to $2^{o(k)}$, even when $n$ is $O(k)$.

However, for planar graphs, and more generally, for graphs excluding some fixed graph as a minor, a vertex cover of size $k$ can be found in time $2^{O(\sqrt{k})} n^{O(1)}$, i.e., the problem is subexponential fixed-parameter tractable. This algorithm is again optimal, in the sense that, under the exponential time hypothesis, no algorithm can solve vertex cover on planar graphs in time $2^{o(\sqrt{k})} n^{O(1)}$.

## Approximate Evaluation

One can find a factor-2 approximation by repeatedly taking *both* endpoints of an edge into the vertex cover, then removing them from the graph. Put otherwise, we find a maximal matching $M$ with a greedy algorithm and construct a vertex cover $C$ that consists of all endpoints of the edges in $M$. In the following figure, a maximal matching $M$ is marked with red, and the vertex cover $C$ is marked with blue.

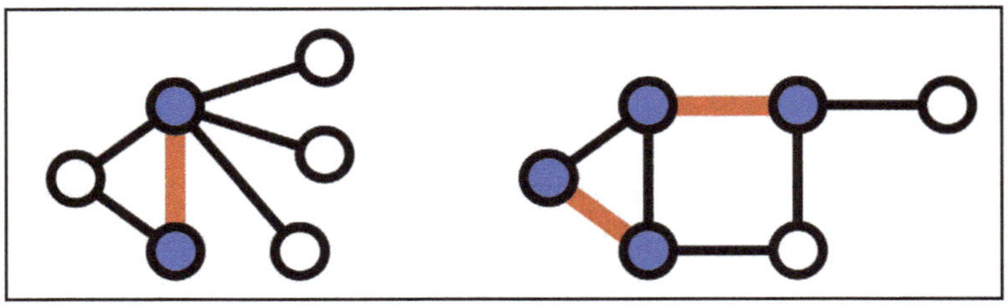

The set $C$ constructed this way is a vertex cover: suppose that an edge $e$ is not covered by $C$; then $M \cup \{e\}$ is a matching and $e \notin M$, which is a contradiction with the assumption that $M$ is maximal. Furthermore, if $e = \{u, v\} \in M$, then any vertex cover – including an optimal vertex cover – must contain $u$ or $v$ (or both); otherwise the edge $e$ is not covered. That is, an optimal cover contains at least *one* endpoint of each edge in $M$; in total, the set $C$ is at most 2 times as large as the optimal vertex cover.

This simple algorithm was discovered independently by Fanica Gavril and Mihalis Yannakakis.

More involved techniques show that there are approximation algorithms with a slightly better approximation factor. For example, an approximation algorithm with an approximation factor of $2-\Theta\left(1/\sqrt{\log|V|}\right)$ is known. The problem can be approximated with an approximation factor $2/(1+\delta)$ in $\delta$ - dense graphs.

## Inapproximability

No better constant-factor approximation algorithm than the above one is known. The minimum vertex cover problem is APX-complete, that is, it cannot be approximated arbitrarily well unless P = NP. Using techniques from the PCP theorem, Dinur and Safra proved in 2005 that minimum vertex cover cannot be approximated within a factor of 1.3606 for any sufficiently large vertex degree unless P = NP. Moreover, if the unique games conjecture is true then minimum vertex cover cannot be approximated within any constant factor better than 2.

Although finding the minimum-size vertex cover is equivalent to finding the maximum-size independent set, as described above, the two problems are not equivalent in an approximation-preserving way: The Independent Set problem has *no* constant-factor approximation unless P = NP.

## Pseudo Code

```
APPROXIMATION-VERTEX-COVER(G):
C = ∅
```

```
E′= G.E
while E′≠ ∅:
  let (u, v) be an arbitrary edge of E′
  C = C ∪ {u, v}
  remove from E′ every edge incident on either u or v
return C
```

## Vertex Cover in Hypergraphs

Given a collection of sets, a set which intersects all sets in the collection in at least one element is called a hitting set, and a simple NP-hard problem is to find a hitting set of smallest size or minimum hitting set. By mapping the sets in the collection onto hyperedges, this can be understood as a natural generalization of the vertex cover problem to hypergraphs which is often just called vertex cover for hypergraphs and, in a more combinatorial context, transversal. The notions of hitting set and set cover are equivalent.

Formally, let $H = (V, E)$ be a hypergraph with vertex set $V$ and hyperedge set $E$. Then a set $S \subseteq V$ is called *hitting set* of $H$ if, for all edges $e \in E$, it holds $S \cap e \neq \emptyset$. The computational problems *minimum hitting set* and *hitting set* are defined as in the case of graphs. Note that we get back the case of vertex covers for simple graphs if the maximum size of the hyperedges is 2.

If that size is restricted to $d$, the problem of finding a minimum $d$-hitting set permits a $d$-approximation algorithm. Assuming the unique games conjecture, this is the best constant-factor algorithm that is possible and otherwise there is the possibility of improving the approximation to $d - 1$.

### Fixed-parameter Tractability

For the hitting set problem, different parametrizations make sense. The hitting set problem is $W$-complete for the parameter OPT, that is, it is unlikely that there is an algorithm that runs in time $f(\mathrm{OPT})n^{O(1)}$ where OPT is the cardinality of the smallest hitting set. The hitting set problem is fixed-parameter tractable for the parameter OPT + $d$, where $d$ is the size of the largest edge of the hypergraph. More specifically, there is an algorithm for hitting set that runs in time $d^{\mathrm{OPT}}n^{O(1)}$.

### Hitting Set and Set Cover

The hitting set problem is equivalent to the set cover problem: An instance of set cover can be viewed as an arbitrary bipartite graph, with sets represented by vertices on the left, elements of the universe represented by vertices on the right, and edges representing the inclusion of elements in sets. The task is then to find a minimum cardinality subset of left-vertices which covers all of the right-vertices. In the hitting set problem, the objective is to cover the left-vertices using a minimum subset of the right vertices. Converting from one problem to the other is therefore achieved by interchanging the two sets of vertices.

### Applications

An example of a practical application involving the hitting set problem arises in efficient dynamic detection of race conditions. In this case, each time global memory is written, the current thread and set of locks held by that thread are stored. Under lockset-based detection, if later another thread writes to that location and there is *not* a race, it must be because it holds at least one lock in common with each of the previous writes. Thus the size of the hitting set represents the minimum lock set size to be race-free. This is useful in eliminating redundant write events, since large lock sets are considered unlikely in practice.

## Universal Vertex

In graph theory, a universal vertex is a vertex of an undirected graph that is adjacent to all other vertices of the graph. It may also be called a dominating vertex, as it forms a one-element dominating set in the graph.

A graph that contains a universal vertex may be called a cone. In this context, the universal vertex may also be called the apex of the cone. However, this terminology conflicts with the terminology of apex graphs, in which an apex is a vertex whose removal leaves a planar subgraph.

### In Special Families of Graphs

The stars are exactly the trees that have a universal vertex, and may be constructed by adding a universal vertex to an independent set. The wheel graphs, similarly, may be formed by adding a universal vertex to a cycle graph. In geometry, the three-dimensional pyramids have wheel graphs as their skeletons, and more generally the graph of any higher-dimensional pyramid has a universal vertex as the apex of the pyramid.

The trivially perfect graphs (the comparability graphs of order-theoretic trees) always contain a universal vertex, the root of the tree, and more strongly they may be characterized as the graphs in which every connected induced subgraph contains a universal vertex. The connected threshold graphs form a subclass of the trivially perfect graphs, so they also contain a universal vertex; they may be defined as the graphs that can be formed by repeated addition of either a universal vertex or an isolated vertex (one with no incident edges).

Every graph with a universal vertex is a dismantlable graph, and almost all dismantlable graphs have a universal vertex.

### Other Properties

In a graph with $n$ vertices, a universal vertex is a vertex whose degree is exactly $n - 1$. Therefore, like the split graphs, graphs with a universal vertex can be recognized purely by their degree sequences, without looking at the structure of the graph.

# Graph Property

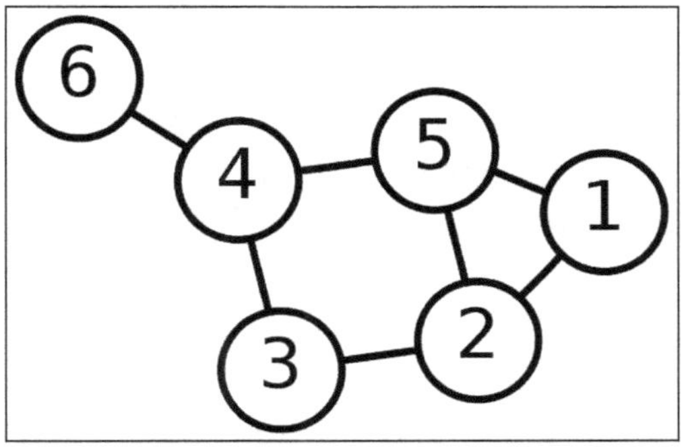


An example graph, with the properties of being planar and being connected, and with order 6, size 7, diameter 3, girth 3, vertex connectivity 1, and degree sequence <3, 3, 3, 2, 2, 1>

In graph theory, a graph property or graph invariant is a property of graphs that depends only on the abstract structure, not on graph representations such as particular labellings or drawings of the graph.

While graph drawing and graph representation are valid topics in graph theory, in order to focus only on the abstract structure of graphs, a graph property is defined to be a property preserved under all possible isomorphisms of a graph. In other words, it is a property of the graph itself, not of a specific drawing or representation of the graph. Informally, the term "graph invariant" is used for properties expressed quantitatively, while "property" usually refers to descriptive characterizations of graphs. For example, the statement "graph does not have vertices of degree 1" is a "property" while "the number of vertices of degree 1 in a graph" is an "invariant".

More formally, a graph property is a class of graphs with the property that any two isomorphic graphs either both belong to the class, or both do not belong to it. Equivalently, a graph property may be formalized using the indicator function of the class, a function from graphs to Boolean values that is true for graphs in the class and false otherwise; again, any two isomorphic graphs must have the same function value as each other. A graph invariant or graph parameter may similarly be formalized as a function from graphs to a broader class of values, such as integers, real numbers, sequences of numbers, or polynomials, that again has the same value for any two isomorphic graphs.

## Properties of Properties

Many graph properties are well-behaved with respect to certain natural partial orders or preorders defined on graphs:

- A graph property $P$ is *hereditary* if every induced subgraph of a graph with property $P$ also has property $P$. For instance, being a perfect graph or being a chordal graph are hereditary properties.
- A graph property is *monotone* if every subgraph of a graph with property $P$ also has property $P$. For instance, being a bipartite graph or being a triangle-free graph is monotone. Every monotone property is hereditary, but not necessarily vice versa; for instance, subgraphs of chordal graphs are not necessarily chordal, so being a chordal graph is not monotone.

- A graph property is *minor-closed* if every graph minor of a graph with property *P* also has property *P*. For instance, being a planar graph is minor-closed. Every minor-closed property is monotone, but not necessarily vice versa; for instance, minors of triangle-free graphs are not necessarily themselves triangle-free.

These definitions may be extended from properties to numerical invariants of graphs: a graph invariant is hereditary, monotone, or minor-closed if the function formalizing the invariant forms a monotonic function from the corresponding partial order on graphs to the real numbers.

Additionally, graph invariants have been studied with respect to their behavior with regard to disjoint unions of graphs:

- A graph invariant is *additive* if, for all two graphs *G* and *H*, the value of the invariant on the disjoint union of *G* and *H* is the sum of the values on *G* and on *H*. For instance, the number of vertices is additive.
- A graph invariant is *multiplicative* if, for all two graphs *G* and *H*, the value of the invariant on the disjoint union of *G* and *H* is the product of the values on *G* and on *H*. For instance, the Hosoya index (number of matchings) is multiplicative.
- A graph invariant is *maxing* if, for all two graphs *G* and *H*, the value of the invariant on the disjoint union of *G* and *H* is the maximum of the values on *G* and on *H*. For instance, the chromatic number is maxing.

In addition, graph properties can be classified according to the type of graph they describe: whether the graph is undirected or directed, whether the property applies to multigraphs, etc.

## Values of Invariants

The target set of a function that defines a graph invariant may be one of:

- A truth-value, true or false, for the indicator function of a graph property.
- An integer, such as the number of vertices or chromatic number of a graph.
- A real number, such as the fractional chromatic number of a graph.
- A sequence of integers, such as the degree sequence of a graph.
- A polynomial, such as the Tutte polynomial of a graph.

## Graph Invariants and Graph Isomorphism

Easily computable graph invariants are instrumental for fast recognition of graph isomorphism, or rather non-isomorphism, since for any invariant at all, two graphs with different values cannot (by definition) be isomorphic. Two graphs with the same invariants may or may not be isomorphic, however.

A graph invariant $I(G)$ is called complete if the identity of the invariants $I(G)$ and $I(H)$ implies the isomorphism of the graphs *G* and *H*. Finding an efficiently-computable such invariant (the problem of graph canonization) would imply an easy solution to the challenging graph isomorphism problem. However, even polynomial-valued invariants such as the chromatic polynomial are not usually complete. The claw graph and the path graph on 4 vertices both have the same chromatic polynomial, for example.

# Graph Factorization

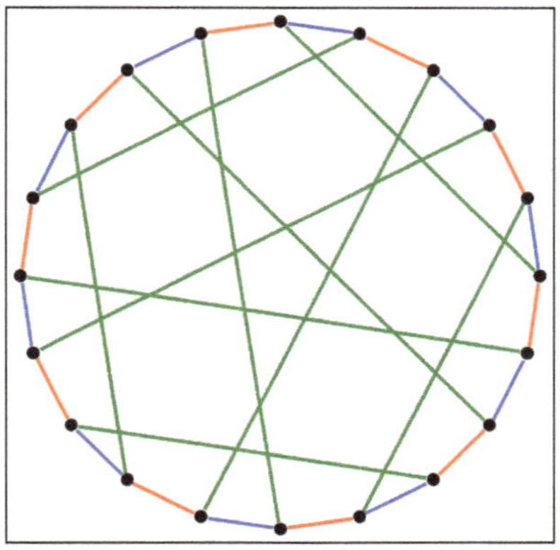

1-factorization of Desargues graph: each color class is a 1-factor.

In graph theory, a factor of a graph *G* is a spanning subgraph, i.e., a subgraph that has the same vertex set as *G*. A *k*-factor of a graph is a spanning *k*-regular subgraph, and a *k*-factorization partitions the edges of the graph into disjoint *k*-factors. A graph *G* is said to be *k*-factorable if it admits a *k*-factorization. In particular, a 1-factor is a perfect matching, and a 1-factorization of a *k*-regular graph is an edge coloring with *k* colors. A 2-factor is a collection of cycles that spans all vertices of the graph.

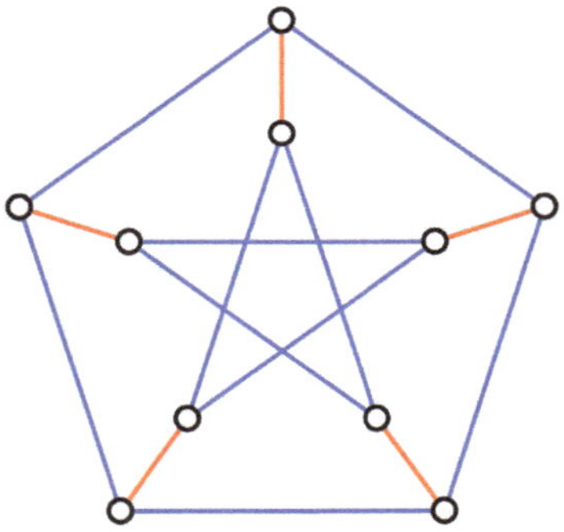

Petersen graph can be partitioned into a 1-factor (red) and a 2-factor (blue). However, the graph is not 1-factorable.

## 1-factorization

If a graph is 1-factorable (ie, has a 1-factorization), then it has to be a regular graph. However, not all regular graphs are 1-factorable. A *k*-regular graph is 1-factorable if it has chromatic index *k*; examples of such graphs include:

- Any regular bipartite graph. Hall's marriage theorem can be used to show that a *k*-regular bipartite graph contains a perfect matching. One can then remove the perfect matching to obtain a $(k - 1)$-regular bipartite graph, and apply the same reasoning repeatedly.

- Any complete graph with an even number of nodes.

However, there are also $k$-regular graphs that have chromatic index $k + 1$, and these graphs are not 1-factorable; examples of such graphs include:

- Any regular graph with an odd number of nodes.
- The Petersen graph.

## Complete Graphs

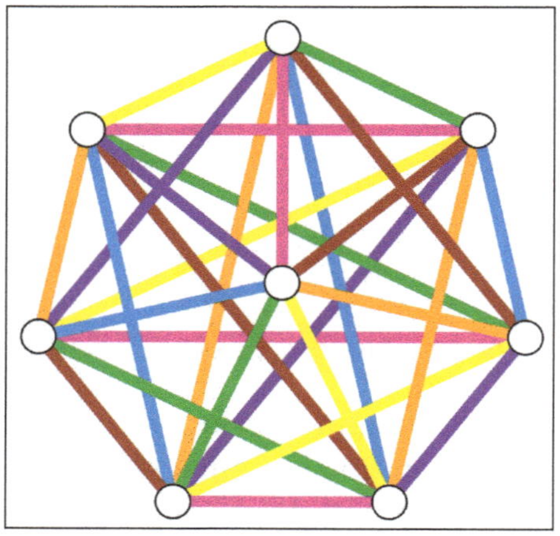

1-factorization of $K_8$ in which each 1-factor consists of an edge from the center to a vertex of a heptagon together with all possible perpendicular edges.

A 1-factorization of a complete graph corresponds to pairings in a round-robin tournament. The 1-factorization of complete graphs is a special case of Baranyai's theorem concerning the 1-factorization of complete hypergraphs.

One method for constructing a 1-factorization of a complete graph on an even number of vertices involves placing all but one of the vertices on a circle, forming a regular polygon, with the remaining vertex at the center of the circle. With this arrangement of vertices, one way of constructing a 1-factor of the graph is to choose an edge $e$ from the center to a single polygon vertex together with all possible edges that lie on lines perpendicular to $e$. The 1-factors that can be constructed in this way form a 1-factorization of the graph.

The number of distinct 1-factorizations of $K_2$, $K_4$, $K_6$, $K_8$, ... is 1, 1, 6, 6240, 1225566720, 252282619805368320, 98758655816833727741338583040, ... OEIS: A000438.

## 1-factorization Conjecture

Let $G$ be a $k$-regular graph with $2n$ nodes. If $k$ is sufficiently large, it is known that $G$ has to be 1-factorable:

- If $k = 2n - 1$, then $G$ is the complete graph $K_{2n}$, and hence 1-factorable.

- If $k = 2n - 2$, then $G$ can be constructed by removing a perfect matching from $K_{2n}$. Again, $G$ is 1-factorable.
- Chetwynd & Hilton (1985) show that if $k \geq 12n/7$, then $G$ is 1-factorable.

The 1-factorization conjecture is a long-standing conjecture that states that $k \approx n$ is sufficient. In precise terms, the conjecture is:

- If $n$ is odd and $k \geq n$, then $G$ is 1-factorable. If $n$ is even and $k \geq n - 1$ then $G$ is 1-factorable.

The overfull conjecture implies the 1-factorization conjecture.

## Perfect 1-factorization

A perfect pair from a 1-factorization is a pair of 1-factors whose union induces a Hamiltonian cycle.

A perfect 1-factorization (P1F) of a graph is a 1-factorization having the property that every pair of 1-factors is a perfect pair. A perfect 1-factorization should not be confused with a perfect matching (also called a 1-factor).

In 1964, Anton Kotzig conjectured that every complete graph $K_{2n}$ where $n \geq 2$ has a perfect 1-factorization. So far, it is known that the following graphs have a perfect 1-factorization:

- The infinite family of complete graphs $K_{2p}$ where $p$ is an odd prime (by Anderson and also Nakamura, independently),
- The infinite family of complete graphs $K_{p+1}$ where $p$ is an odd prime,
- Sporadic additional results, including $K_{2n}$ where $2n \in \{16, 28, 36, 40, 50, 126, 170, 244, 344, 730, 1332, 1370, 1850, 2198, 3126, 6860, 12168, 16808, 29792\}$. Some newer results are collected here.

If the complete graph $K_{n+1}$ has a perfect 1-factorization, then the complete bipartite graph $K_{n,n}$ also has a perfect 1-factorization.

## 2-factorization

If a graph is 2-factorable, then it has to be $2k$-regular for some integer $k$. Julius Petersen showed in 1891 that this necessary condition is also sufficient: any $2k$-regular graph is 2-factorable.

If a connected graph is $2k$-regular and has an even number of edges it may also be $k$-factored, by choosing each of the two factors to be an alternating subset of the edges of an Euler tour. This applies only to connected graphs; disconnected counterexamples include disjoint unions of odd cycles, or of copies of $K_{2k+1}$.

The Oberwolfach problem concerns the existence of 2-factorizations of complete graphs into isomorphic subgraphs. It asks for which subgraphs this is possible. This is known when the subgraph is connected (in which case it is a Hamiltonian cycle and this special case is the problem of Hamiltonian decomposition) but the general case remains unsolved.

# Graph Isomorphism

In graph theory, an isomorphism of graphs $G$ and $H$ is a bijection between the vertex sets of $G$ and $H$,

$$f: V(G) \to V(H)$$

such that any two vertices $u$ and $v$ of $G$ are adjacent in $G$ if and only if $f(u)$ and $f(v)$ are adjacent in $H$. This kind of bijection is commonly described as "edge-preserving bijection", in accordance with the general notion of isomorphism being a structure-preserving bijection.

If an isomorphism exists between two graphs, then the graphs are called isomorphic and denoted as $G \simeq H$. In the case when the bijection is a mapping of a graph onto itself, i.e., when $G$ and $H$ are one and the same graph, the bijection is called an automorphism of $G$.

Graph isomorphism is an equivalence relation on graphs and as such it partitions the class of all graphs into equivalence classes. A set of graphs isomorphic to each other is called an isomorphism class of graphs.

The two graphs shown below are isomorphic, despite their different looking drawings.

| Graph G | Graph H | An isomorphism between G and H |
| --- | --- | --- |
| 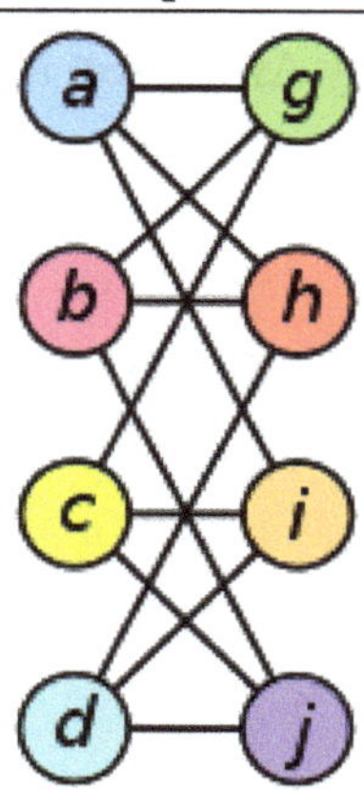 | 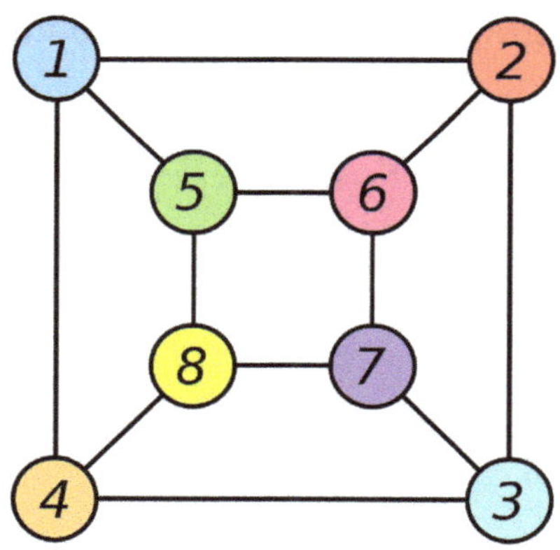 | f(a) = 1<br>f(b) = 6<br>f(c) = 8<br>f(d) = 3<br>f(g) = 5<br>f(h) = 2<br>f(i) = 4 |

## Variations

Graphs are understood to be uni-directed non-labeled non-weighted graphs. However, the notion of isomorphic may be applied to all other variants of the notion of graph, by adding the requirements to preserve the corresponding additional elements of structure: arc directions, edge weights, etc., with the following exception.

## Isomorphism of Labeled Graphs

For labeled graphs, two definitions of isomorphism are in use.

Under one definition, an isomorphism is a vertex bijection which is both edge-preserving and label-preserving.

Under another definition, an isomorphism is an edge-preserving vertex bijection which preserves equivalence classes of labels, i.e., vertices with equivalent (e.g., the same) labels are mapped onto the vertices with equivalent labels and vice versa; same with edge labels.

For example, the $K_2$ graph with the two vertices labelled with 1 and 2 has a single automorphism under the first definition, but under the second definition there are two auto-morphisms.

The second definition is assumed in certain situations when graphs are endowed with *unique labels* commonly taken from the integer range 1,...,$n$, where $n$ is the number of the vertices of the graph, used only to uniquely identify the vertices. In such cases two labeled graphs are sometimes said to be isomorphic if the corresponding underlying unlabeled graphs are isomorphic (otherwise the definition of isomorphism would be trivial).

The formal notion of "isomorphism", e.g., of "graph isomorphism", captures the informal notion that some objects have "the same structure" if one ignores individual distinctions of "atomic" components of objects in question. Whenever individuality of "atomic" components (vertices and edges, for graphs) is important for correct representation of whatever is modeled by graphs, the model is refined by imposing additional restrictions on the structure, and other mathematical objects are used: digraphs, labeled graphs, colored graphs, rooted trees and so on. The isomorphism relation may also be defined for all these generalizations of graphs: the isomorphism bijection must preserve the elements of structure which define the object type in question: arcs, labels, vertex/edge colors, the root of the rooted tree, etc.

The notion of "graph isomorphism" allows us to distinguish graph properties inherent to the structures of graphs themselves from properties associated with graph representations: graph drawings, data structures for graphs, graph labelings, etc. For example, if a graph has exactly one cycle, then all graphs in its isomorphism class also have exactly one cycle. On the other hand, in the common case when the vertices of a graph are (*represented* by) the integers 1, 2,... $N$, then the expression,

$$\sum_{v \in V(G)} v \cdot \deg v$$

may be different for two isomorphic graphs.

## Whitney Theorem

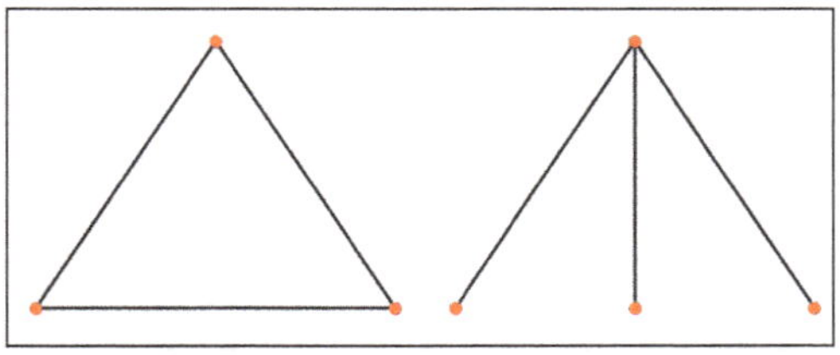

The exception of Whitney's theorem: these two graphs are not isomorphic but have isomorphic line graphs.

The Whitney graph isomorphism theorem, shown by Hassler Whitney, states that two connected graphs are isomorphic if and only if their line graphs are isomorphic, with a single exception: $K_3$, the complete graph on three vertices, and the complete bipartite graph $K_{1,3}$, which are not isomorphic but both have $K_3$ as their line graph. The Whitney graph theorem can be extended to hypergraphs.

### Recognition of Graph Isomorphism

While graph isomorphism may be studied in a classical mathematical way, as exemplified by the Whitney theorem, it is recognized that it is a problem to be tackled with an algorithmic approach. The computational problem of determining whether two finite graphs are isomorphic is called the graph isomorphism problem.

Its practical applications include primarily cheminformatics, mathematical chemistry (identification of chemical compounds), and electronic design automation (verification of equivalence of various representations of the design of an electronic circuit).

The graph isomorphism problem is one of few standard problems in computational complexity theory belonging to NP, but not known to belong to either of its well-known (and, if P ≠ NP, disjoint) subsets: P and NP-complete. It is one of only two, out of 12 total, problems listed in Garey & Johnson (1979) whose complexity remains unresolved, the other being integer factorization. It is however known that if the problem is NP-complete then the polynomial hierarchy collapses to a finite level.

In November 2015, László Babai, a mathematician and computer scientist at the University of Chicago, claimed to have proven that the graph isomorphism problem is solvable in quasi-polynomial time. This work has not yet been vetted. In January 2017, Babai briefly retracted the quasi-polynomiality claim and stated a sub-exponential time time complexity bound instead. He restored the original claim five days later.

Its generalization, the subgraph isomorphism problem, is known to be NP-complete.

The main areas of research for the problem are design of fast algorithms and theoretical investigations of its computational complexity, both for the general problem and for special classes of graphs.

## Graph Homomorphism

In the mathematical field of graph theory, a graph homomorphism is a mapping between two graphs that respects their structure. More concretely, it is a function between the vertex sets of two graphs that maps adjacent vertices to adjacent vertices.

Homomorphisms generalize various notions of graph colorings and allow the expression of an important class of constraint satisfaction problems, such as certain scheduling or frequency assignment problems. The fact that homomorphisms can be composed leads to rich algebraic structures: a preorder on graphs, a distributive lattice, and a category (one for undirected graphs and one for directed graphs). The computational complexity of finding a homomorphism between given graphs is prohibitive in general, but a lot is known about special cases that are solvable in polynomial time. Boundaries between tractable and intractable cases have been an active area of research.

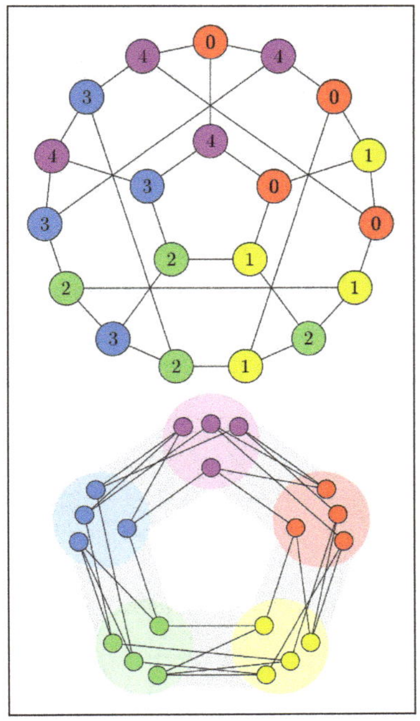

A homomorphism from the flower snark $J_5$ into the cycle graph $C_5$. It is also a retraction onto the subgraph on the central five vertices. Thus $J_5$ is in fact homomorphically equivalent to the core $C_5$.

*Graphs* are finite, undirected graphs with loops allowed, but multiple edges (parallel edges) disallowed. A graph homomorphism $f$ from a graph $G = (V(G), E(G))$ to a graph $H = (V(H), E(H))$, written,

$$f: G \to H$$

is a function from $V(G)$ to $V(H)$ that maps endpoints of each edge in $G$ to endpoints of an edge in $H$. Formally, $\{u,v\} \in E(G)$ implies $\{f(u),f(v)\} \in E(H)$, for all pairs of vertices $u$, $v$ in $V(G)$. If there exists any homomorphism from $G$ to $H$, then $G$ is said to be homomorphic to $H$ or $H$-colorable. This is often denoted as just:

$$G \to H.$$

The above definition is extended to directed graphs. Then, for a homomorphism $f: G \to H$, $(f(u),f(v))$ is an arc (directed edge) of $H$ whenever $(u,v)$ is an arc of $G$.

There is an injective homomorphism from $G$ to $H$ (i.e., one that never maps distinct vertices to one vertex) if and only if $G$ is a subgraph of $H$. If a homomorphism $f: G \to H$ is a bijection (a one-to-one correspondence between vertices of $G$ and $H$) whose inverse function is also a graph homomorphism, then $f$ is a graph isomorphism.

Covering maps are a special kind of homomorphisms that mirror the definition and many properties of covering maps in topology. They are defined as surjective homomorphisms (i.e., something maps to each vertex) that are also locally bijective, that is, a bijection on the neighbourhood of each vertex. An example is the bipartite double cover, formed from a graph by splitting each vertex $v$ into $v_0$ and $v_1$ and replacing each edge $u,v$ with edges $u_0,v_1$ and $v_0,u_1$. The function mapping $v_0$ and $v_1$ in the cover to $v$ in the original graph is a homomorphism and a covering map.

Graph homeomorphism is a different notion, not related directly to homomorphisms. Roughly speaking, it requires injectivity, but allows mapping edges to paths (not just to edges). Graph minors are a still more relaxed notion.

## Cores and Retracts

Two graphs $G$ and $H$ are homomorphically equivalent if $G \to H$ and $H \to G$. The maps are not necessarily surjective or nor injective. For instance, the complete bipartite graphs $K_{2,2}$ and $K_{3,3}$ are homomorphically equivalent: each map can be defined as taking the left (resp. right) half of the domain graph and mapping to just one vertex in the left (resp. right) half of the image graph.

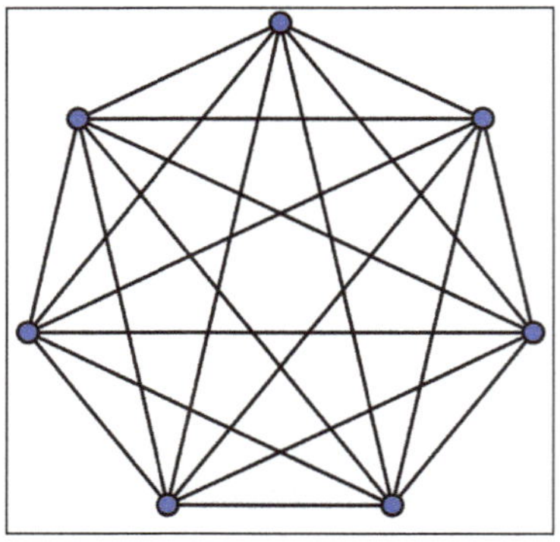

$K_7$, the complete graph with 7 vertices, is a core.

A retraction is a homomorphism $r$ from a graph $G$ to a subgraph $H$ of $G$ such that $r(v) = v$ for each vertex $v$ of $H$. In this case the subgraph $H$ is called a retract of $G$.

A core is a graph with no homomorphism to any proper subgraph. Equivalently, a core can be defined as a graph that does not retract to any proper subgraph. Every graph $G$ is homomorphically equivalent to a unique core (up to isomorphism), called *the core* of $G$. Notably, this is not true in general for infinite graphs. However, the same definitions apply to directed graphs and a directed graph is also equivalent to a unique core. Every graph and every directed graph contains its core as a retract and as an induced subgraph.

For example, all complete graphs $K_n$ and all odd cycles (cycle graphs of odd length) are cores. Every 3-colorable graph $G$ that contains a triangle (that is, has the complete graph $K_3$ as a subgraph) is homomorphically equivalent to $K_3$. This is because, on one hand, a 3-coloring of $G$ is the same as a homomorphism $G \to K_3$, On the other hand, every subgraph of $G$ trivially admits a homomorphism into $G$, implying $K_3 \to G$. This also means that $K_3$ is the core of any such graph $G$. Similarly, every bipartite graph that has at least one edge is equivalent to $K_2$.

## Connection to Colorings

A $k$-coloring, for some integer $k$, is an assignment of one of $k$ colors to each vertex of a graph $G$ such that the endpoints of each edge get different colors. The $k$-colorings of $G$ correspond exactly to homomorphisms from $G$ to the complete graph $K_k$. Indeed, the vertices of $K_k$ correspond to the $k$ colors, and two colors are adjacent as vertices of $K_k$ if and only if they are different. Hence a function defines a homomorphism to $K_k$ if and only if it maps adjacent vertices of $G$ to different colors (i.e., it is a $k$-coloring). In particular, $G$ is $k$-colorable if and only if it is $K_k$-colorable.

If there are two homomorphisms $G \to H$ and $H \to K_k$, then their composition $G \to K_k$ is also a homomorphism. In other words, if a graph $H$ can be colored with $k$ colors, and there is a homomorphism from $G$ to $H$, then $G$ can also be $k$-colored. Therefore, $G \to H$ implies $\chi(G) \leq \chi(H)$, where $\chi$ denotes the chromatic number of a graph (the least $k$ for which it is $k$-colorable).

## Variants

General homomorphisms can also be thought of as a kind of coloring: if the vertices of a fixed graph $H$ are the available *colors* and edges of $H$ describe which colors are *compatible*, then an $H$-coloring of $G$ is an assignment of colors to vertices of $G$ such that adjacent vertices get compatible colors. Many notions of graph coloring fit into this pattern and can be expressed as graph homomorphisms into different families of graphs. Circular colorings can be defined using homomorphisms into circular complete graphs, refining the usual notion of colorings. Fractional and $b$-fold coloring can be defined using homomorphisms into Kneser graphs. T-colorings correspond to homomorphisms into certain infinite graphs. An oriented coloring of a directed graph is a homomorphism into any oriented graph. An L(2,1)-coloring is a homomorphism into the complement of the path graph that is locally injective, meaning it is required to be injective on the neighbourhood of every vertex.

## Orientations without Long Paths

Another interesting connection concerns orientations of graphs. An orientation of an undirected graph $G$ is any directed graph obtained by choosing one of the two possible orientations for each edge. An example of an orientation of the complete graph $K_k$ is the transitive tournament $\vec{T}_k$ with vertices 1,2,…,$k$ and arcs from $i$ to $j$ whenever $i < j$. A homomorphism between orientations of graphs $G$ and $H$ yields a homomorphism between the undirected graphs $G$ and $H$, simply by disregarding the orientations. On the other hand, given a homomorphism $G \to H$ between undirected graphs, any orientation $\vec{H}$ of $H$ can be pulled back to an orientation $\vec{G}$ of $G$ so that $\vec{G}$ has a homomorphism to $\vec{H}$. Therefore, a graph $G$ is $k$-colorable (has a homomorphism to $K_k$) if and only if some orientation of $G$ has a homomorphism to $\vec{T}_k$.

A folklore theorem states that for all $k$, a directed graph $G$ has a homomorphism to $T\!\to_k$ if and only if it admits no homomorphism from the directed path $P\!\to_{k+1}$. Here $P\!\to_n$ is the directed graph with vertices 1, 2, …, $n$ and edges from $i$ to $i + 1$, for $i = 1, 2, …, n - 1$. Therefore, a graph is $k$-colorable if and only if it has an orientation that admits no homomorphism from $P\!\to_{k+1}$. This statement can be strengthened slightly to say that a graph is $k$-colorable if and only if some orientation contains no directed path of length $k$ (no $P\!\to_{k+1}$ as a subgraph). This is the Gallai–Hasse–Roy–Vitaver theorem.

## Connection to Constraint Satisfaction Problems

## Examples

Some scheduling problems can be modeled as a question about finding graph homomorphisms. As an example, one might want to assign workshop courses to time slots in a calendar so that two courses attended by the same student are not too close to each other in time. The courses form a graph $G$, with an edge between any two courses that are attended by some common student. The time slots form a graph $H$, with an edge between any two slots that are distant enough in time. For instance, if one wants a cyclical, weekly schedule, such that each student gets their workshop courses on non-consecutive days, then $H$ would be the complement graph of $C_7$. A graph homomorphism from $G$ to $H$ is then a schedule assigning courses to time slots, as specified. To add a requirement saying that, e.g., no single student has courses on both Friday and Monday, it suffices to remove the corresponding edge from $H$.

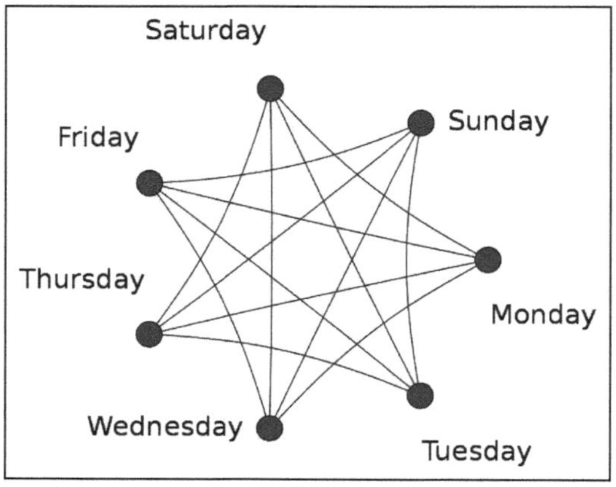


Graph $H$ of non-consecutive weekdays, isomorphic to the complement graph of $C_7$ and to the circular clique $K_{7/2}$.

A simple frequency allocation problem can be specified as follows: a number of transmitters in a wireless network must choose a frequency channel on which they will transmit data. To avoid interference, transmitters that are geographically close should use channels with frequencies that are far apart. If this condition is approximated with a single threshold to define 'geographically close' and 'far apart', then a valid channel choice again corresponds to a graph homomorphism. It should go from the graph of transmitters $G$, with edges between pairs that are geographically close, to the graph of channels $H$, with edges between channels that are far apart. While this model is rather simplified, it does admit some flexibility: transmitter pairs that are not close but could interfere because of geographical features can added to the edges of $G$. Those that do not communicate at the same time can be removed from it. Similarly, channel pairs that are far apart but exhibit harmonic interference can be removed from the edge set of $H$.

In each case, these simplified models display many of the issues that have to be handled in practice. Constraint satisfaction problems, which generalize graph homomorphism problems, can express various additional types of conditions (such as individual preferences, or bounds on the number of coinciding assignments). This allows the models to be made more realistic and practical.

## Formal View

Graphs and directed graphs can be viewed as a special case of the far more general notion called relational structures (defined as a set with a tuple of relations on it). Directed graphs are structures with a single binary relation (adjacency) on the domain (the vertex set). Under this view, homomorphisms of such structures are exactly graph homomorphisms. In general, the question of finding a homomorphism from one relational structure to another is a constraint satisfaction problem (CSP). The case of graphs gives a concrete first step that helps to understand more complicated CSPs. Many algorithmic methods for finding graph homomorphisms, like backtracking, constraint propagation and local search, apply to all CSPs.

For graphs $G$ and $H$, the question of whether $G$ has a homomorphism to $H$ corresponds to a CSP instance with only one kind of constraint, as follows. The *variables* are the vertices of $G$ and the *domain* for each variable is the vertex set of $H$. An *evaluation* is a function that assigns to each variable an element of the domain, so a function $f$ from $V(G)$ to $V(H)$. Each edge or arc $(u,v)$ of $G$ then corresponds to the *constraint* $((u,v), E(H))$. This is a constraint expressing that the evaluation should map the arc $(u,v)$ to a pair $(f(u),f(v))$ that is in the relation $E(H)$, that is, to an arc of $H$. A solution to the CSP is an evaluation that respects all constraints, so it is exactly a homomorphism from $G$ to $H$.

## Structure of Homomorphisms

Compositions of homomorphisms are homomorphisms. In particular, the relation $\to$ on graphs is transitive (and reflexive, trivially), so it is a preorder on graphs. Let the equivalence class of a graph $G$ under homomorphic equivalence be $[G]$. The equivalence class can also be represented by the unique core in $[G]$. The relation $\to$ is a partial order on those equivalence classes; it defines a poset.

Let $G < H$ denote that there is a homomorphism from $G$ to $H$, but no homomorphism from $H$ to $G$. The relation $\to$ is a dense order, meaning that for all (undirected) graphs $G$, $H$ such that $G < H$, there is a graph $K$ such that $G < K < H$ (this holds except for the trivial cases $G = K_0$ or $K_1$). For example, between any two complete graphs (except $K_0$, $K_1$) there are infinitely many circular complete graphs, corresponding to rational numbers between natural numbers.

The poset of equivalence classes of graphs under homomorphisms is a distributive lattice, with the join of $[G]$ and $[H]$ defined as (the equivalence class of) the disjoint union $[G \cup H]$, and the meet of $[G]$ and $[H]$ defined as the tensor product $[G \times H]$ (the choice of graphs $G$ and $H$ representing the equivalence classes $[G]$ and $[H]$ does not matter). The join-irreducible elements of this lattice are exactly connected graphs. This can be shown using the fact that a homomorphism maps a connected graph into one connected component of the target graph. The meet-irreducible elements of this lattice are exactly the multiplicative graphs. These are the graphs $K$ such that a product $G \times H$ has a homomorphism to $K$ only when one of $G$ or $H$ also does. Identifying multiplicative graphs lies at the heart of Hedetniemi's conjecture.

Graph homomorphisms also form a category, with graphs as objects and homomorphisms as arrows. The initial object is the empty graph, while the terminal object is the graph with one vertex and one loop at that vertex. The tensor product of graphs is the category-theoretic product and the exponential graph is the exponential object for this category. Since these two operations are always defined, the category of graphs is a cartesian closed category. For the same reason, the lattice of equivalence classes of graphs under homomorphisms is in fact a Heyting algebra.

For directed graphs the same definitions apply. In particular $\to$ is a partial order on equivalence classes of directed graphs. It is distinct from the order $\to$ on equivalence classes of undirected graphs, but contains it as a suborder. This is because every undirected graph can be thought of as a directed graph where every arc $(u,v)$ appears together with its inverse arc $(v,u)$, and this does not change the definition of homomorphism. The order $\to$ for directed graphs is again a distributive lattice and a Heyting algebra, with join and meet operations defined as before. However, it is not dense. There is also a category with directed graphs as objects and homomorphisms as arrows, which is again a cartesian closed category.

## Incomparable Graphs

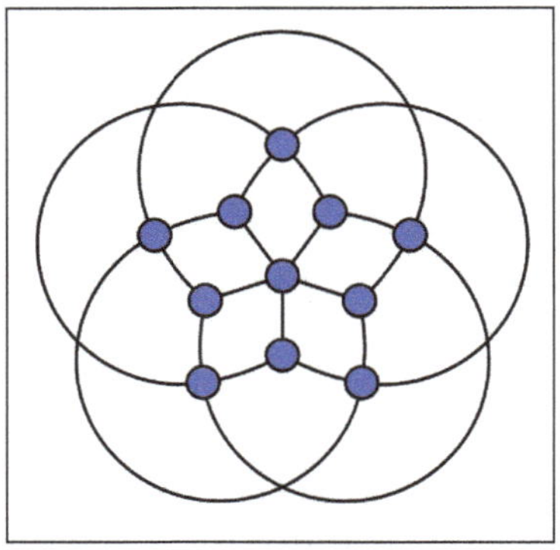

The Grötzsch graph, incomparable to $K_3$.

There are many incomparable graphs with respect to the homomorphism preorder, that is, pairs of graphs such that neither admits a homomorphism into the other. One way to construct them is to consider the odd girth of a graph *G*, the length of its shortest odd-length cycle. The odd girth is, equivalently, the smallest odd number *g* for which there exists a homomorphism from the cycle graph on *g* vertices to *G*. For this reason, if $G \to H$, then the odd girth of *G* is greater than or equal to the odd girth of *H*.

On the other hand, if $G \to H$, then the chromatic number of *G* is less than or equal to the chromatic number of *H*. Therefore, if *G* has strictly larger odd girth than *H* and strictly larger chromatic number than *H*, then *G* and *H* are incomparable. For example, the Grötzsch graph is 4-chromatic and triangle-free (it has girth 4 and odd girth 5), so it is incomparable to the triangle graph $K_3$.

Examples of graphs with arbitrarily large values of odd girth and chromatic number are Kneser graphs and generalized Mycielskians. A sequence of such graphs, with simultaneously increasing values of both parameters, gives infinitely many incomparable graphs (an antichain in the homomorphism preorder). Other properties, such as density of the homomorphism preorder, can be proved using such families. Constructions of graphs with large values of chromatic number and girth, not just odd girth, are also possible, but more complicated.

Among directed graphs, it is much easier to find incomparable pairs. For example, consider the directed cycle graphs $\vec{C}_n$, with vertices 1, 2, ..., *n* and edges from *i* to *i* + 1 (for *i* = 1, 2, ..., *n* – 1) and from *n* to 1. There is a homomorphism from $\vec{C}_n$ to $\vec{C}_k$ ($n, k \geq 3$) if and only if *n* is a multiple of *k*. In particular, directed cycle graphs $\vec{C}_n$ with *n* prime are all incomparable.

## Computational Complexity

In the graph homomorphism problem, an instance is a pair of graphs (*G*,*H*) and a solution is a homomorphism from *G* to *H*. The general decision problem, asking whether there is any solution, is NP-complete. However, limiting allowed instances gives rise to a variety of different problems, some of which are much easier to solve. Methods that apply when restraining the left side *G* are very different than for the right side *H*, but in each case a dichotomy (a sharp boundary between easy and hard cases) is known or conjectured.

## Homomorphisms to a Fixed Graph

The homomorphism problem with a fixed graph $H$ on the right side of each instance is also called the $H$-coloring problem. When $H$ is the complete graph $K_k$, this is the graph $k$-coloring problem, which is solvable in polynomial time for $k$ = 0, 1, 2, and NP-complete otherwise. In particular, $K_2$-colorability of a graph $G$ is equivalent to $G$ being bipartite, which can be tested in linear time. More generally, whenever $H$ is a bipartite graph, $H$-colorability is equivalent to $K_2$-colorability (or $K_0$ / $K_1$-colorability when $H$ is empty/edgeless), hence equally easy to decide. Pavol Hell and Jaroslav Nešetřil proved that, for undirected graphs, no other case is tractable:

> "Hell–Nešetřil theorem: The $H$-coloring problem is in P when $H$ is bipartite and NP-complete otherwise".

This is also known as the *dichotomy theorem* for (undirected) graph homomorphisms, since it divides $H$-coloring problems into NP-complete or P problems, with no intermediate cases. For directed graphs, the situation is more complicated and in fact equivalent to the much more general question of characterizing the complexity of constraint satisfaction problems. It turns out that $H$-coloring problems for directed graphs are just as general and as diverse as CSPs with any other kinds of constraints. Formally, a (finite) *constraint language* (or *template*) $\Gamma$ is a finite domain and a finite set of relations over this domain. CSP($\Gamma$) is the constraint satisfaction problem where instances are only allowed to use constraints in $\Gamma$.

> "Theorem: For every constraint language $\Gamma$, the problem CSP($\Gamma$) is equivalent under polynomial-time reductions to some $H$-coloring problem, for some directed graph $H$".

Intuitively, this means that every algorithmic technique or complexity result that applies to $H$-coloring problems for directed graphs $H$ applies just as well to general CSPs. In particular, one can ask whether the Hell–Nešetřil theorem can be extended to directed graphs. By the above theorem, this is equivalent to the Feder–Vardi conjecture (aka CSP conjecture, dichotomy conjecture) on CSP dichotomy, which states that for every constraint language $\Gamma$, CSP($\Gamma$) is NP-complete or in P. This conjecture was proved in 2017 independently by Dmitry Zhuk and Andrei Bulatov, leading the following corollary:

> "Corollary: The $H$-coloring problem on directed graphs, for a fixed $H$, is either in P or NP-complete".

## Homomorphisms from a Fixed Family of Graphs

The homomorphism problem with a single fixed graph $G$ on left side of input instances can be solved by brute-force in time $|V(H)|^{O(|V(G)|)}$, so polynomial in the size of the input graph $H$. In other words, the problem is trivially in P for graphs $G$ of bounded size. The interesting question is then what other properties of $G$, beside size, make polynomial algorithms possible.

The crucial property turns out to be treewidth, a measure of how tree-like the graph is. For a graph $G$ of treewidth at most $k$ and a graph $H$, the homomorphism problem can be solved in time $|V(H)|^{O(k)}$ with a standard dynamic programming approach. In fact, it is enough to assume that the core of $G$ has treewidth at most $k$. This holds even if the core is not known.

The exponent in the $|V(H)|^{O(k)}$-time algorithm cannot be lowered significantly: no algorithm with running time $|V(H)|^{o(tw(G)/\log tw(G))}$ exists, assuming the exponential time hypothesis (ETH), even if the inputs are restricted to any class of graphs of unbounded treewidth. The ETH is an unproven assumption similar to P ≠ NP, but stronger. Under the same assumption, there are also essentially no other properties that can be used to get polynomial time algorithms. This is formalized as follows:

> "Theorem (Grohe): For a computable class of graphs $\mathcal{G}$, the homomorphism problem for instances $(G,H)$ with $G \in \mathcal{G}$ is in P if and only if graphs in $\mathcal{G}$ have cores of bounded treewidth (assuming ETH)".

One can ask whether the problem is at least solvable in a time arbitrarily highly dependent on *G*, but with a fixed polynomial dependency on the size of *H*. The answer is again positive if we limit *G* to a class of graphs with cores of bounded treewidth, and negative for every other class. In the language of parameterized complexity, this formally states that the homomorphism problem in $\mathcal{G}$ parameterized by the size (number of edges) of *G* exhibits a dichotomy. It is fixed-parameter tractable if graphs in $\mathcal{G}$ have cores of bounded treewidth, and W-complete otherwise.

The same statements hold more generally for constraint satisfaction problems (or for relational structures, in other words). The only assumption needed is that constraints can involve only a bounded number of variables (all relations are of some bounded arity, 2 in the case of graphs). The relevant parameter is then the treewidth of the primal constraint graph.

# Graph Automorphism

In the mathematical field of graph theory, an automorphism of a graph is a form of symmetry in which the graph is mapped onto itself while preserving the edge–vertex connectivity.

Formally, an automorphism of a graph $G = (V,E)$ is a permutation σ of the vertex set *V*, such that the pair of vertices $(u,v)$ form an edge if and only if the pair $(\sigma(u),\sigma(v))$ also form an edge. That is, it is a graph isomorphism from *G* to itself. Automorphisms may be defined in this way both for directed graphs and for undirected graphs. The composition of two automorphisms is another automorphism, and the set of automorphisms of a given graph, under the composition operation, forms a group, the automorphism group of the graph. In the opposite direction, by Frucht's theorem, all groups can be represented as the automorphism group of a connected graph – indeed, of a cubic graph.

## Computational Complexity

Constructing the automorphism group is at least as difficult (in terms of its computational complexity) as solving the graph isomorphism problem, determining whether two given graphs correspond vertex-for-vertex and edge-for-edge. For, *G* and *H* are isomorphic if and only if the disconnected graph formed by the disjoint union of graphs *G* and *H* has an automorphism that swaps the two components. In fact, just counting the automorphisms is polynomial-time equivalent to graph isomorphism.

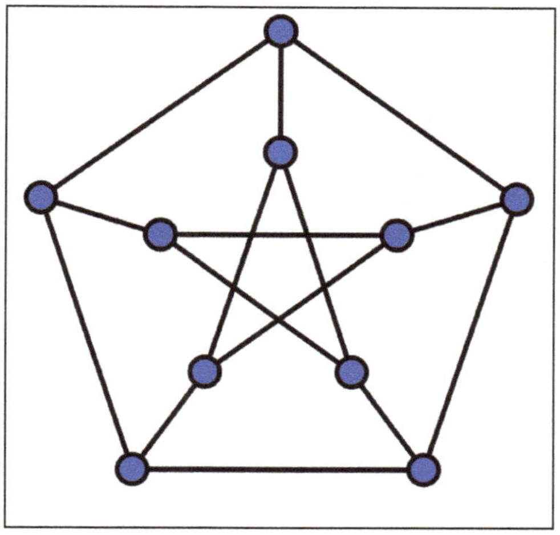

This drawing of the Petersen graph displays a subgroup of its symmetries, isomorphic to the dihedral group $D_5$, but the graph has additional symmetries that are not present in the drawing. For example, since the graph is symmetric, all edges are equivalent.

The graph automorphism problem is the problem of testing whether a graph has a nontrivial automorphism. It belongs to the class NP of computational complexity. Similar to the graph isomorphism problem, it is unknown whether it has a polynomial time algorithm or it is NP-complete. There is a polynomial time algorithm for solving the graph automorphism problem for graphs where vertex degrees are bounded by a constant. The graph automorphism problem is polynomial-time many-one reducible to the graph isomorphism problem, but the converse reduction is unknown. By contrast, hardness is known when the automorphisms are constrained in a certain fashion; for instance, determining the existence of a fixed-point-free automorphism (an automorphism that fixes no vertex) is NP-complete, and the problem of counting such automorphisms is P-complete.

## Algorithms, Software and Applications

While no *worst-case* polynomial-time algorithms are known for the general Graph Automorphism problem, finding the automorphism group (and printing out an irredundant set of generators) for many large graphs arising in applications is rather easy. Several open-source software tools are available for this task, including NAUTY, BLISS and SAUCY. SAUCY and BLISS are particularly efficient for sparse graphs, e.g., SAUCY processes some graphs with millions of vertices in mere seconds. However, BLISS and NAUTY can also produce Canonical Labeling, whereas SAUCY is currently optimized for solving Graph Automorphism. An important observation is that for a graph on $n$ vertices, the automorphism group can be specified by no more than $n-1$ generators, and the above software packages are guaranteed to satisfy this bound as a side-effect of their algorithms (minimal sets of generators are harder to find and are not particularly useful in practice). It also appears that the total support (i.e., the number of vertices moved) of all generators is limited by a linear function of $n$, which is important in runtime analysis of these algorithms. However, this has not been established for a fact, as of March 2012.

Practical applications of Graph Automorphism include graph drawing and other visualization tasks, solving structured instances of Boolean Satisfiability arising in the context of Formal verification and Logistics. Molecular symmetry can predict or explain chemical properties.

## Symmetry Display

Several graph drawing researchers have investigated algorithms for drawing graphs in such a way that the automorphisms of the graph become visible as symmetries of the drawing. This may be done either by using a method that is not designed around symmetries, but that automatically generates symmetric drawings when possible, or by explicitly identifying symmetries and using them to guide vertex placement in the drawing. It is not always possible to display all symmetries of the graph simultaneously, so it may be necessary to choose which symmetries to display and which to leave unvisualized.

## Graph Families Defined by their Automorphisms

Several families of graphs are defined by having certain types of automorphisms:

- An asymmetric graph is an undirected graph with only the trivial automorphism.
- A vertex-transitive graph is an undirected graph in which every vertex may be mapped by an automorphism into any other vertex.
- An edge-transitive graph is an undirected graph in which every edge may be mapped by an automorphism into any other edge.
- A symmetric graph is a graph such that every pair of adjacent vertices may be mapped by an automorphism into any other pair of adjacent vertices.
- A distance-transitive graph is a graph such that every pair of vertices may be mapped by an automorphism into any other pair of vertices that are the same distance apart.
- A semi-symmetric graph is a graph that is edge-transitive but not vertex-transitive.
- A half-transitive graph is a graph that is vertex-transitive and edge-transitive but not symmetric.
- A skew-symmetric graph is a directed graph together with a permutation σ on the vertices that maps edges to edges but reverses the direction of each edge. Additionally, σ is required to be an involution.

Inclusion relationships between these families are indicated by the following table:

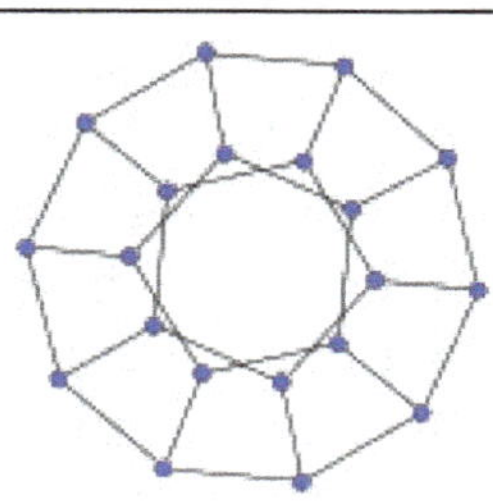

distance-transitive →

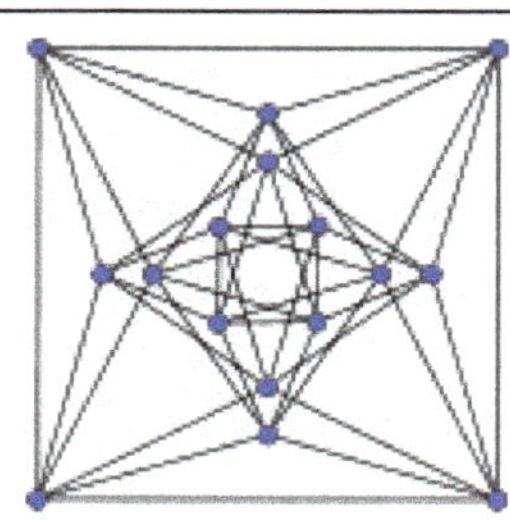

distance-regular ←

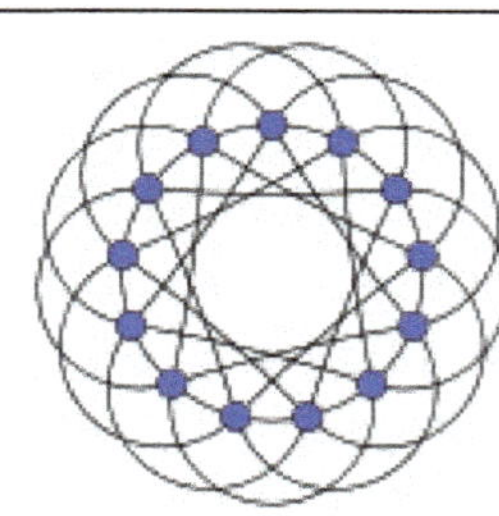

strongly regular

↓

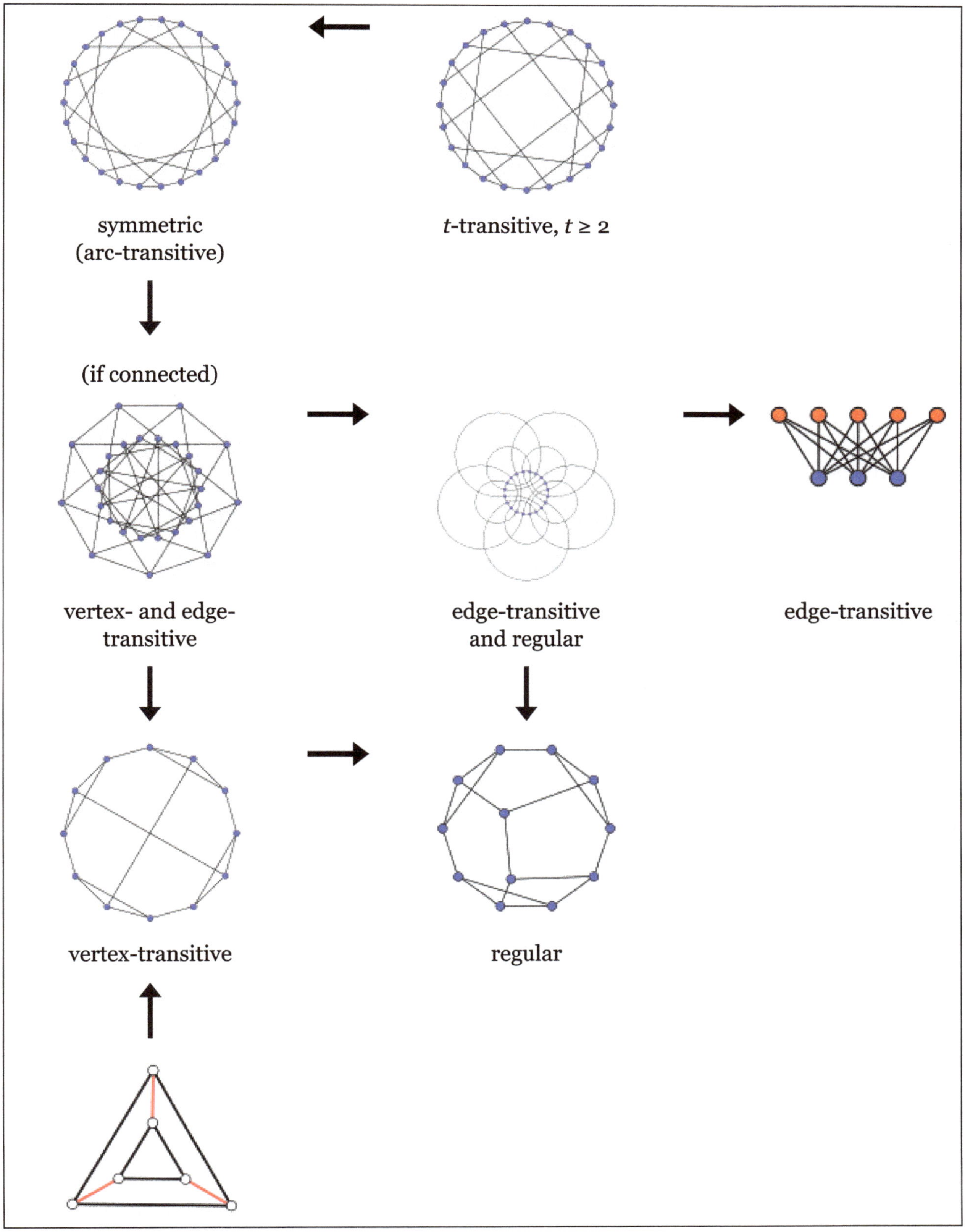
symmetric
(arc-transitive)
t-transitive, t ≥ 2
(if connected)
vertex- and edge-
transitive
edge-transitive
and regular
edge-transitive
vertex-transitive
regular

# Graph Amalgamation

In graph theory, a graph amalgamation is a relationship between two graphs (one graph is an amalgamation of another). Similar relationships include subgraphs and minors. Amalgamations can provide a way to reduce a graph to a simpler graph while keeping certain structure intact. The amalgamation can then be used to study properties of the original graph in an easier to understand context. Applications include embeddings, computing genus distribution, and Hamiltonian decompositions.

Let $G$ and $H$ be two graphs with the same number of edges where $G$ has more vertices than $H$. Then we say that $H$ is an amalgamation of $G$ if there is a bijection $\phi: E(G) \to E(H)$ and a surjection $\psi: V(G) \to V(H)$ and the following hold:

- If $x$, $y$ are two vertices in $G$ where $\psi(x) \neq \psi(y)$, and both $x$ and $y$ are adjacent by edge $e$ in $G$, then $\psi(x)$ and $\psi(y)$ are adjacent by edge $\phi(e)$ in $H$.
- If $e$ is a loop on a vertex $x \in V(G)$, then $\phi(e)$ is a loop on $\psi(x) \in H$.
- If $e$ joins $x, y \in V(G)$, where $x \neq y$, but $\psi(x) = \psi(y)$, then $\phi(e)$ is a loop on $\psi(x)$.

Note that while $G$ can be a graph or a pseudograph, it will usually be the case that $H$ is a pseudograph.

## Properties

Edge colorings are invariant to amalgamation. This is obvious, as all of the edges between the two graphs are in bijection with each other. However, what may not be obvious, is that if   is a complete graph of the form $K_{2n+1}$, and we color the edges as to specify a Hamiltonian decomposition (a decomposition into Hamiltonian paths, then those edges also form a Hamiltonian Decomposition in $H$).

## Example

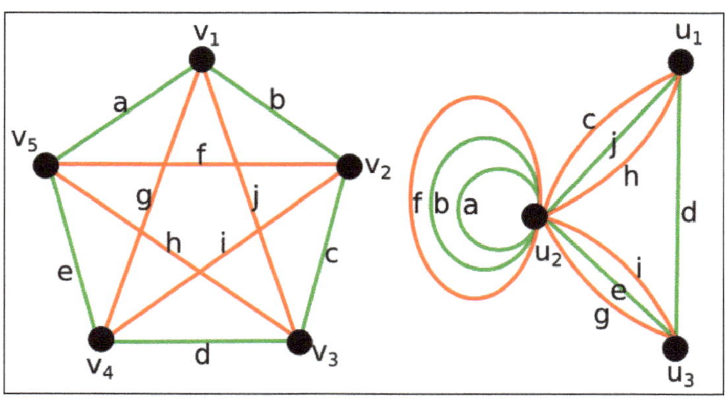


An amalgamation of the complete graph on five vertices.

Figure illustrates an amalgamation of $K_5$. The invariance of edge coloring and Hamiltonian Decomposition can be seen clearly. The function $\phi$ is a bijection and is given as letters in the figure.

The function $\psi$ is given in the table below:

| $v \in V(G)$ | $\psi(v)$ |
|---|---|
| $v_1$ | $u_2$ |
| $v_2$ | $u_2$ |
| $v_3$ | $u_1$ |
| $v_4$ | $u_3$ |
| $v_5$ | $u_2$ |

### Hamiltonian Decompositions

One of the ways in which amalgamations can be used is to find Hamiltonian Decompositions of complete graphs with 2*n* + 1 vertices. The idea is to take a graph and produce an amalgamation of it which is edge colored in $n$ colors and satisfies certain properties (called an outline Hamiltonian decomposition). We can then 'reverse' the amalgamation and we are left with $K_{2n+1}$ colored in a Hamiltonian Decomposition.

In Hilton outlines a method for doing this, as well as a method for finding all Hamiltonian Decompositions without repetition. The methods rely on a theorem he provides which states (roughly) that if we have an outline Hamiltonian decomposition, we could have arrived at it by first starting with a Hamiltonian decomposition of the complete graph and then finding an amalgamation for it.

## Vertex-transitive Graph

In the mathematical field of graph theory, a vertex-transitive graph is a graph *G* such that, given any two vertices $v_1$ and $v_2$ of *G*, there is some automorphism,

$$f: V(G) \to V(G)$$

such that:

$$f(v_1) = v_2.$$

In other words, a graph is vertex-transitive if its automorphism group acts transitively upon its vertices. A graph is vertex-transitive if and only if its graph complement is, since the group actions are identical.

Every symmetric graph without isolated vertices is vertex-transitive, and every vertex-transitive graph is regular. However, not all vertex-transitive graphs are symmetric (for example, the edges of the truncated tetrahedron), and not all regular graphs are vertex-transitive (for example, the Frucht graph and Tietze's graph).

## Finite Examples

Finite vertex-transitive graphs include the symmetric graphs (such as the Petersen graph, the Heawood graph and the vertices and edges of the Platonic solids). The finite Cayley graphs (such as cube-connected cycles) are also vertex-transitive, as are the vertices and edges of the Archimedean solids (though only two of these are symmetric). Potočnik, Spiga and Verret have constructed a census of all connected cubic vertex-transitive graphs on at most 1280 vertices.

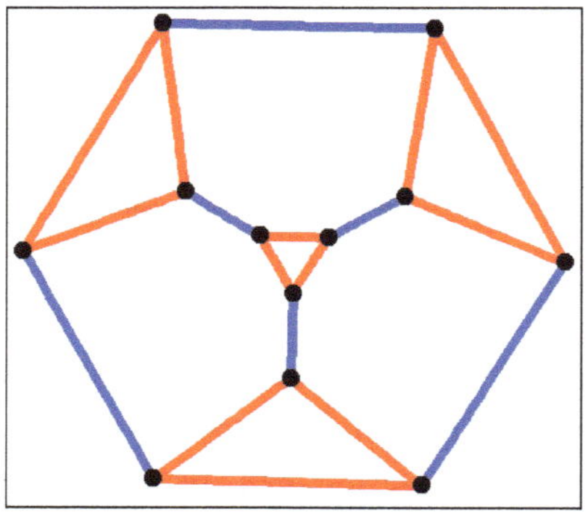

The edges of the truncated tetrahedron form a vertex-transitive graph (also a Cayley graph) which is not symmetric.

Although every Cayley graph is vertex-transitive, there exist other vertex-transitive graphs that are not Cayley graphs. The most famous example is the Petersen graph, but others can be constructed including the line graphs of edge-transitive non-bipartite graphs with odd vertex degrees.

## Properties

The edge-connectivity of a vertex-transitive graph is equal to the degree $d$, while the vertex-connectivity will be at least $2(d + 1)/3$. If the degree is 4 or less, or the graph is also edge-transitive, or the graph is a minimal Cayley graph, then the vertex-connectivity will also be equal to $d$.

## Infinite Examples

Infinite vertex-transitive graphs include:

- Infinite paths (infinite in both directions).
- Infinite regular trees, e.g. The cayley graph of the free group.
- Graphs of uniform tessellations, including all tilings by regular polygons.
- Infinite cayley graphs.
- The rado graph.

Two countable vertex-transitive graphs are called quasi-isometric if the ratio of their distance functions is bounded from below and from above. A well known conjecture stated that every infinite vertex-transitive graph is quasi-isometric to a Cayley graph. A counterexample was proposed by Diestel and Leader in 2001. In 2005, Eskin, Fisher, and Whyte confirmed the counterexample.

# Regular Graph

In graph theory, a regular graph is a graph where each vertex has the same number of neighbors; i.e. every vertex has the same degree or valency. A regular directed graph must also satisfy the stronger condition that the indegree and outdegree of each vertex are equal to each other. A regular graph with vertices of degree *k* is called a *k*regular graph or regular graph of degree *k*. Also, from the handshaking lemma, a regular graph of odd degree will contain an even number of vertices.

Regular graphs of degree at most 2 are easy to classify: A 0-regular graph consists of disconnected vertices, a 1-regular graph consists of disconnected edges, and a 2-regular graph consists of a disjoint union of cycles and infinite chains.

A 3-regular graph is known as a cubic graph.

A strongly regular graph is a regular graph where every adjacent pair of vertices has the same number *l* of neighbors in common, and every non-adjacent pair of vertices has the same number *n* of neighbors in common. The smallest graphs that are regular but not strongly regular are the cycle graph and the circulant graph on 6 vertices.

The complete graph $K_m$ is strongly regular for any $m$.

A theorem by Nash-Williams says that every *k*regular graph on 2*k* + 1 vertices has a Hamiltonian cycle.

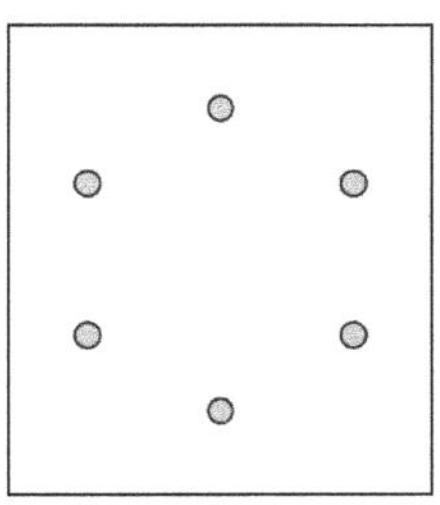
0-regular graph.

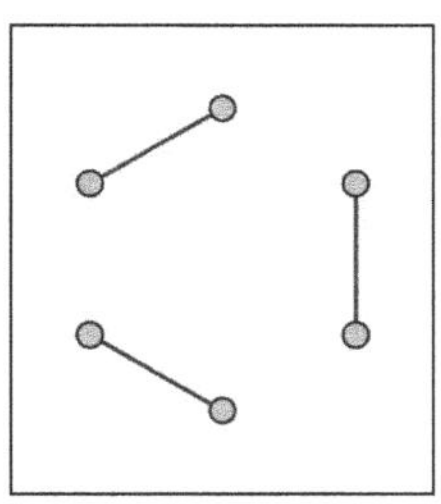
1-regular graph.

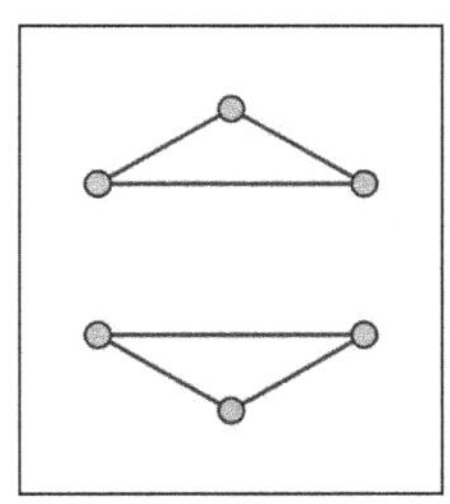
2-regular graph.

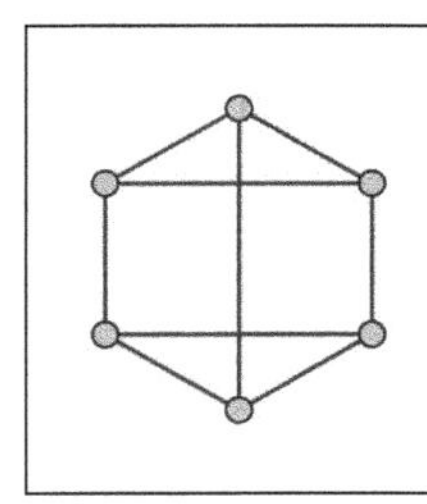
3-regular graph.

## Existence

It is well known that the necessary and sufficient conditions for a $k$ regular graph of order $n$ to exist are that $n \geq k+1$ and that $nk$ is even. In such case it is easy to construct regular graphs by considering appropriate parameters for circulant graphs.

## Algebraic Properties

Let *A* be the adjacency matrix of a graph. Then the graph is regular if and only if $\mathrm{j} = (1,\ldots,1)$ is an eigenvector of *A*. Its eigenvalue will be the constant degree of the graph. Eigenvectors corresponding to other eigenvalues are orthogonal to $\mathrm{j}$, so for such eigenvectors $v = (v_1,\ldots,v_n)$, we have $\sum_{i=1}^{n} v_i = 0$.

A regular graph of degree *k* is connected if and only if the eigenvalue *k* has multiplicity one. The "only if" direction is a consequence of the Perron–Frobenius theorem.

There is also a criterion for regular and connected graphs : a graph is connected and regular if and only if the matrix of ones $J$, with $J_{ij}=1$, is in the adjacency algebra of the graph (meaning it is a linear combination of powers of $A$).

Let $G$ be a $k$-regular graph with diameter $D$ and eigenvalues of adjacency matrix $k=\lambda_0>\lambda_1\geq\cdots\geq\lambda_{n-1}$. If $G$ is not bipartite, then:

$$D\leq\frac{\log(n-1)}{\log(\lambda_0/\lambda_1)}+1$$

### Generation

Fast algorithms exist to enumerate, up to isomorphism, all regular graphs with a given degree and number of vertices.

## Planar Graph

In graph theory, a planar graph is a graph that can be embedded in the plane, i.e., it can be drawn on the plane in such a way that its edges intersect only at their endpoints. In other words, it can be drawn in such a way that no edges cross each other. Such a drawing is called a plane graph or planar embedding of the graph. A plane graph can be defined as a planar graph with a mapping from every node to a point on a plane, and from every edge to a plane curve on that plane, such that the extreme points of each curve are the points mapped from its end nodes, and all curves are disjoint except on their extreme points.

Every graph that can be drawn on a plane can be drawn on the sphere as well, and vice versa, by means of stereographic projection.

Plane graphs can be encoded by combinatorial maps.

The equivalence class of topologically equivalent drawings on the sphere is called a planar map. Although a plane graph has an external or unbounded face, none of the faces of a planar map have a particular status.

Planar graphs generalize to graphs drawable on a surface of a given genus. In this terminology, planar graphs have graph genus 0, since the plane (and the sphere) are surfaces of genus 0.

| Example graphs | |
|---|---|
| Planar | Nonplanar |
| 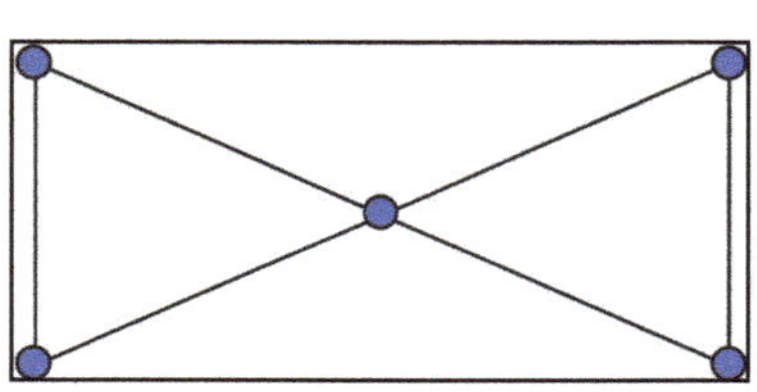 | 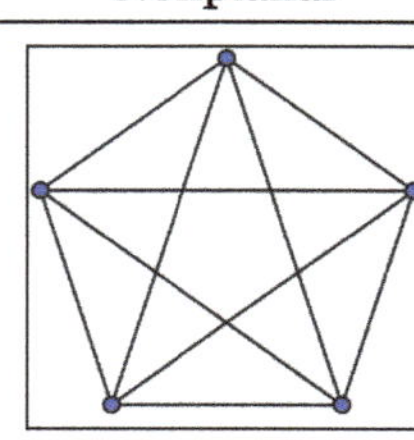 |
| Butterfly graph. | Complete graph $K_5$. |

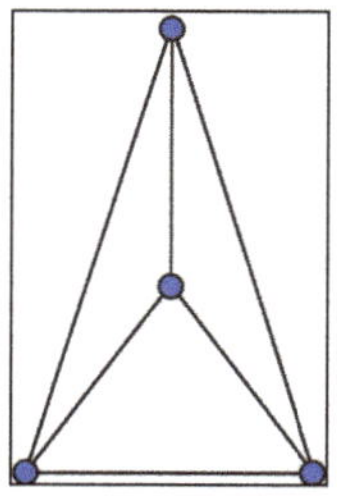

Complete graph $K_4$.

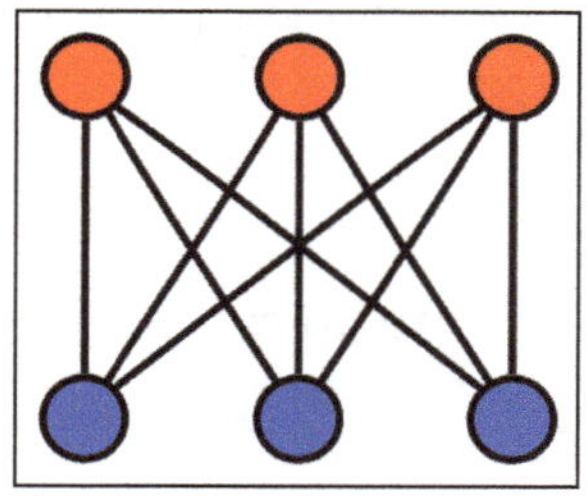

Utility graph $K_{3,3}$.

## Kuratowski's and Wagner's Theorems

The Polish mathematician Kazimierz Kuratowski provided a characterization of planar graphs in terms of forbidden graphs, now known as Kuratowski's theorem:

> A finite graph is planar if and only if it does not contain a subgraph that is a subdivision of the complete graph $K_5$ or the complete bipartite graph $K_{3,3}$ (utility graph).

A subdivision of a graph results from inserting vertices into edges (for example, changing an edge •——• to •—•—•) zero or more times.

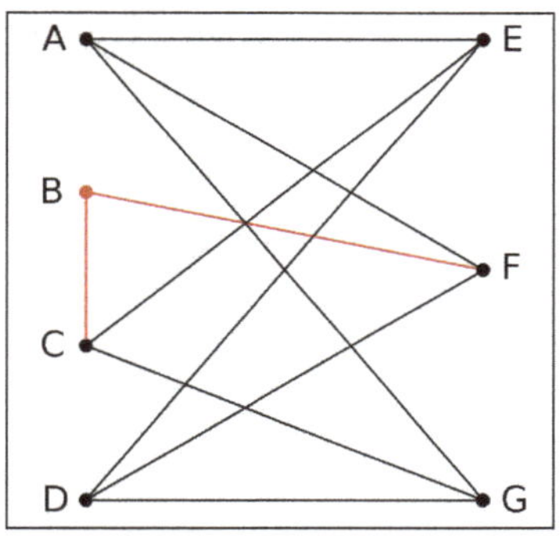


An example of a graph with no $K_5$ or $K_{3,3}$ subgraph. However, it contains a subdivision of $K_{3,3}$ and is therefore non-planar.

Instead of considering subdivisions, Wagner's theorem deals with minors:

> "A finite graph is planar if and only if it does not have $K_5$ or $K_{3,3}$ as a minor".

A minor of a graph results from taking a subgraph and repeatedly contracting an edge into a vertex, with each neighbor of the original end-vertices becoming a neighbor of the new vertex.

Klaus Wagner asked more generally whether any minor-closed class of graphs is determined by a finite set of "forbidden minors". This is now the Robertson–Seymour theorem, proved in a long series of papers. In the language of this theorem, $K_5$ and $K_{3,3}$ are the forbidden minors for the class of finite planar graphs.

## Other Planarity Criteria

In practice, it is difficult to use Kuratowski's criterion to quickly decide whether a given graph is planar. However, there exist fast algorithms for this problem: for a graph with $n$ vertices, it is possible to determine in time O($n$) (linear time) whether the graph may be planar or not.

For a simple, connected, planar graph with $v$ vertices and $e$ edges and $f$ faces, the following simple conditions hold for $v \geq 3$:

- Theorem: $e \leq 3v - 6$.
- Theorem: If there are no cycles of length 3, then $e \leq 2v - 4$.
- Theorem: $f \leq 2v - 4$.

In this sense, planar graphs are sparse graphs, in that they have only $O(v)$ edges, asymptotically smaller than the maximum $O(v^2)$. The graph $K_{3,3}$, for example, has 6 vertices, 9 edges, and no cycles of length 3. Therefore, by Theorem 2, it cannot be planar. These theorems provide necessary conditions for planarity that are not sufficient conditions, and therefore can only be used to prove a graph is not planar, not that it is planar. If both theorem 1 and 2 fail, other methods may be used.

- Whitney's planarity criterion gives a characterization based on the existence of an algebraic dual;
- Mac Lane's planarity criterion gives an algebraic characterization of finite planar graphs, via their cycle spaces;
- The Fraysseix–Rosenstiehl planarity criterion gives a characterization based on the existence of a bipartition of the cotree edges of a depth-first search tree. It is central to the left-right planarity testing algorithm;
- Schnyder's theorem gives a characterization of planarity in terms of partial order dimension;
- Colin de Verdière's planarity criterion gives a characterization based on the maximum multiplicity of the second eigenvalue of certain Schrödinger operators defined by the graph;
- The Hanani–Tutte theorem states that a graph is planar if and only if it has a drawing in which each independent pair of edges crosses an even number of times; it can be used to characterize the planar graphs via a system of equations modulo 2.

## Euler's Formula

Euler's formula states that if a finite, connected, planar graph is drawn in the plane without any edge intersections, and $v$ is the number of vertices, $e$ is the number of edges and $f$ is the number of faces (regions bounded by edges, including the outer, infinitely large region), then:

$$v - e + f = 2.$$

As an illustration, in the butterfly graph given above, $v = 5$, $e = 6$ and $f = 3$. In general, if the property holds for all planar graphs of $f$ faces, any change to the graph that creates an additional face while keeping the graph planar would keep $v - e + f$ an invariant. Since the property holds for all graphs with $f = 2$, by mathematical induction it holds for all cases. Euler's formula can also be proved as follows: if the graph isn't a tree, then remove an edge which completes a cycle. This lowers both $e$ and $f$ by one, leaving $v - e + f$ constant. Repeat until the remaining graph is a tree; trees have $v = e + 1$ and $f = 1$, yielding $v - e + f = 2$, i. e., the Euler characteristic is 2.

In a finite, connected, *simple*, planar graph, any face (except possibly the outer one) is bounded by at least three edges and every edge touches at most two faces; using Euler's formula, one can then show that these graphs are *sparse* in the sense that if $v \geq 3$:

$$e \leq 3v - 6.$$

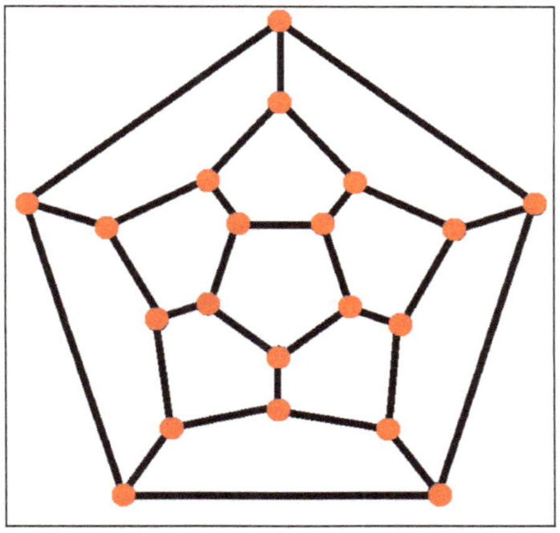

A Schlegel diagram of a regular dodecahedron, forming a planar graph from a convex polyhedron.

Euler's formula is also valid for convex polyhedra. This is no coincidence: every convex polyhedron can be turned into a connected, simple, planar graph by using the Schlegel diagram of the polyhedron, a perspective projection of the polyhedron onto a plane with the center of perspective chosen near the center of one of the polyhedron's faces. Not every planar graph corresponds to a convex polyhedron in this way: the trees do not, for example. Steinitz's theorem says that the polyhedral graphs formed from convex polyhedra are precisely the finite 3-connected simple planar graphs. More generally, Euler's formula applies to any polyhedron whose faces are simple polygons that form a surface topologically equivalent to a sphere, regardless of its convexity.

## Average Degree

Connected planar graphs with more than one edge obey the inequality $2e \geq 3f$, because each face has at least three face-edge incidences and each edge contributes exactly two incidences. It follows via algebraic transformations of this inequality with Euler's formula $v - e + f = 2$ that for finite planar graphs the average degree is strictly less than 6. Graphs with higher average degree cannot be planar.

## Coin Graphs

We say that two circles drawn in a plane *kiss* (or *osculate*) whenever they intersect in exactly one point. A "coin graph" is a graph formed by a set of circles, no two of which have overlapping interiors, by making a vertex for each circle and an edge for each pair of circles that kiss. The circle packing theorem, first proved by Paul Koebe in 1936, states that a graph is planar if and only if it is a coin graph.

This result provides an easy proof of Fáry's theorem, that every planar graph can be embedded in the plane in such a way that its edges are straight line segments that do not cross each other. If one places each vertex of the graph at the center of the corresponding circle in a coin graph representation, then the line segments between centers of kissing circles do not cross any of the other edges.

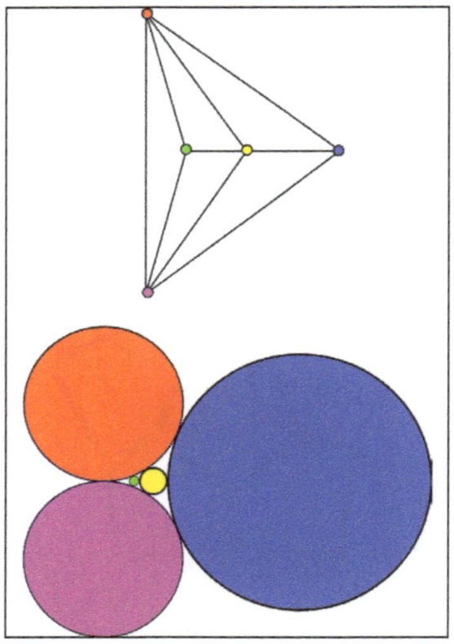

Example of the circle packing theorem on $K^{-}_{5}$, the complete graph on five vertices, minus one edge.

## Planar Graph Density

The density $D$ of a planar graph, or network, is defined as a ratio of the number of edges $E$ to the number of possible edges in a network with $N$ nodes, given by a planar graph $(E_{\max}=3N-6)$, giving $D=\frac{E-N+1}{2N-5}$. A completely sparse planar graph has $D=0$, alternatively a completely dense planar graph has $D=1$.

## Related Families of Graphs

### Maximal Planar Graphs

A simple graph is called maximal planar if it is planar but adding any edge (on the given vertex set) would destroy that property. All faces (including the outer one) are then bounded by three edges, explaining the alternative term plane triangulation. The alternative names "triangular graph" or "triangulated graph" have also been used, but are ambiguous, as they more commonly refer to the line graph of a complete graph and to the chordal graphs respectively. Every maximal planar is 3-vertex-connected.

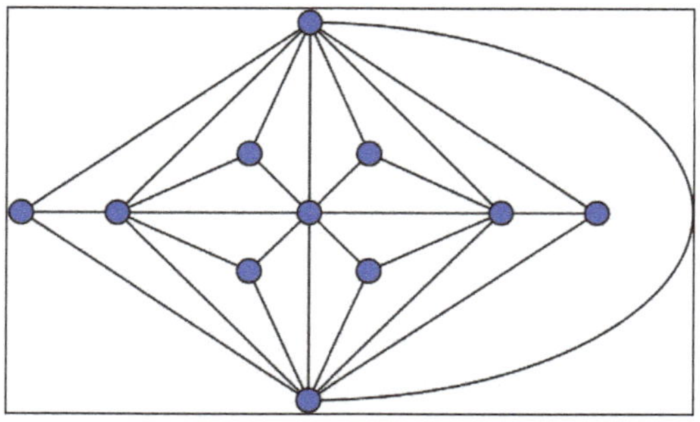

The Goldner–Harary graph is maximal planar. All its faces are bounded by three edges.

If a maximal planar graph has $v$ vertices with $v > 2$, then it has precisely $3v - 6$ edges and $2v - 4$ faces.

Apollonian networks are the maximal planar graphs formed by repeatedly splitting triangular faces into triples of smaller triangles. Equivalently, they are the planar 3-trees.

Strangulated graphs are the graphs in which every peripheral cycle is a triangle. In a maximal planar graph (or more generally a polyhedral graph) the peripheral cycles are the faces, so maximal planar graphs are strangulated. The strangulated graphs include also the chordal graphs, and are exactly the graphs that can be formed by clique-sums (without deleting edges) of complete graphs and maximal planar graphs.

## Outerplanar Graphs

Outerplanar graphs are graphs with an embedding in the plane such that all vertices belong to the unbounded face of the embedding. Every outerplanar graph is planar, but the converse is not true: $K_4$ is planar but not outerplanar. A theorem similar to Kuratowski's states that a finite graph is outerplanar if and only if it does not contain a subdivision of $K_4$ or of $K_{2,3}$. The above is a direct corollary of the fact that a graph $G$ is outerplanar if the graph formed from $G$ by adding a new vertex, with edges connecting it to all the other vertices, is a planar graph.

A 1-outerplanar embedding of a graph is the same as an outerplanar embedding. For $k > 1$ a planar embedding is $k$-outerplanar if removing the vertices on the outer face results in a $(k - 1)$-outerplanar embedding. A graph is $k$-outerplanar if it has a $k$-outerplanar embedding.

## Halin Graphs

A Halin graph is a graph formed from an undirected plane tree (with no degree-two nodes) by connecting its leaves into a cycle, in the order given by the plane embedding of the tree. Equivalently, it is a polyhedral graph in which one face is adjacent to all the others. Every Halin graph is planar. Like outerplanar graphs, Halin graphs have low treewidth, making many algorithmic problems on them more easily solved than in unrestricted planar graphs.

## Other Related Families

An apex graph is a graph that may be made planar by the removal of one vertex, and a $k$-apex graph is a graph that may be made planar by the removal of at most $k$ vertices.

A 1-planar graph is a graph that may be drawn in the plane with at most one simple crossing per edge, and a $k$-planar graph is a graph that may be drawn with at most $k$ simple crossings per edge.

A map graph is a graph formed from a set of finitely many simply-connected interior-disjoint regions in the plane by connecting two regions when they share at least one boundary point. When at most three regions meet at a point, the result is a planar graph, but when four or more regions meet at a point, the result can be nonplanar.

A toroidal graph is a graph that can be embedded without crossings on the torus. More generally, the genus of a graph is the minimum genus of a two-dimensional surface into which the graph may be embedded; planar graphs have genus zero and nonplanar toroidal graphs have genus one.

Any graph may be embedded into three-dimensional space without crossings. However, a three-dimensional analogue of the planar graphs is provided by the linklessly embeddable graphs, graphs that can be embedded into three-dimensional space in such a way that no two cycles are topologically linked with each other. In analogy to Kuratowski's and Wagner's characterizations of the planar graphs as being the graphs that do not contain $K_5$ or $K_{3,3}$ as a minor, the linklessly embeddable graphs may be characterized as the graphs that do not contain as a minor any of the seven graphs in the Petersen family. In analogy to the characterizations of the outerplanar and planar graphs as being the graphs with Colin de Verdière graph invariant at most two or three, the linklessly embeddable graphs are the graphs that have Colin de Verdière invariant at most four.

An upward planar graph is a directed acyclic graph that can be drawn in the plane with its edges as non-crossing curves that are consistently oriented in an upward direction. Not every planar directed acyclic graph is upward planar, and it is NP-complete to test whether a given graph is upward planar.

## Enumeration of Planar Graphs

The asymptotic for the number of (labeled) planar graphs on $n$ vertices is $g \cdot n^{-7/2} \cdot \gamma^n \cdot n!$, where $\gamma \approx 27.22687$ and $g \approx 0.43 \times 10^{-5}$.

Almost all planar graphs have an exponential number of automorphisms.

The number of unlabeled (non-isomorphic) planar graphs on $n$ vertices is between $27.2^n$ and $30.06^n$.

## Other Facts and Definitions

The Four Color Theorem states that every planar graph is 4-colorable (i.e. 4-partite).

Fáry's theorem states that every simple planar graph admits an embedding in the plane such that all edges are straight line segments which don't intersect. A universal point set is a set of points such that every planar graph with $n$ vertices has such an embedding with all vertices in the point set; there exist universal point sets of quadratic size, formed by taking a rectangular subset of the integer lattice. Every simple outerplanar graph admits an embedding in the plane such that all vertices lie on a fixed circle and all edges are straight line segments that lie inside the disk and don't intersect, so $n$-vertex regular polygons are universal for outerplanar graphs.

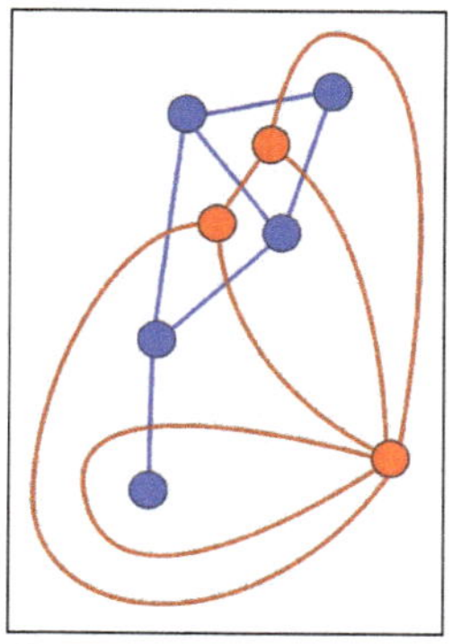

A planar graph and its dual.

Given an embedding $G$ of a (not necessarily simple) connected graph in the plane without edge intersections, we construct the dual graph $G^*$ as follows: we choose one vertex in each face of $G$ (including the outer face) and for each edge $e$ in $G$ we introduce a new edge in $G^*$ connecting the two vertices in $G^*$ corresponding to the two faces in $G$ that meet at $e$. Furthermore, this edge is drawn so that it crosses $e$ exactly once and that no other edge of $G$ or $G^*$ is intersected. Then $G^*$ is again the embedding of a (not necessarily simple) planar graph; it has as many edges as $G$, as many vertices as $G$ has faces and as many faces as $G$ has vertices. The term "dual" is justified by the fact that $G^{**} = G$; here the equality is the equivalence of embeddings on the sphere. If $G$ is the planar graph corresponding to a convex polyhedron, then $G^*$ is the planar graph corresponding to the dual polyhedron.

Duals are useful because many properties of the dual graph are related in simple ways to properties of the original graph, enabling results to be proven about graphs by examining their dual graphs.

While the dual constructed for a particular embedding is unique (up to isomorphism), graphs may have different (i.e. non-isomorphic) duals, obtained from different (i.e. non-homeomorphic) embeddings.

A *Euclidean graph* is a graph in which the vertices represent points in the plane, and the edges are assigned lengths equal to the Euclidean distance between those points.

A plane graph is said to be *convex* if all of its faces (including the outer face) are convex polygons. A planar graph may be drawn convexly if and only if it is a subdivision of a 3-vertex-connected planar graph.

Scheinerman's conjecture (now a theorem) states that every planar graph can be represented as an intersection graph of line segments in the plane.

The planar separator theorem states that every $n$-vertex planar graph can be partitioned into two subgraphs of size at most $2n/3$ by the removal of $O(\sqrt{n})$ vertices. As a consequence, planar graphs also have treewidth and branch-width $O(\sqrt{n})$.

For two planar graphs with $v$ vertices, it is possible to determine in time $O(v)$ whether they are isomorphic or not.

The meshedness coefficient of a planar graph normalizes its number of bounded faces (the same as the circuit rank of the graph, by Mac Lane's planarity criterion) by dividing it by $2n - 5$, the maximum possible number of bounded faces in a planar graph with $n$ vertices. Thus, it ranges from 0 for trees to 1 for maximal planar graphs.

# Induced Subgraph

In graph theory, an induced subgraph of a graph is another graph, formed from a subset of the vertices of the graph and all of the edges connecting pairs of vertices in that subset.

Formally, let $G = (V, E)$ be any graph, and let $S \subset V$ be any subset of vertices of $G$. Then the induced subgraph $G[S]$ is the graph whose vertex set is $S$ and whose edge set consists of all of the edges in $E$

that have both endpoints in $S$. The same definition works for undirected graphs, directed graphs, and even multigraphs.

The induced subgraph $G[S]$ may also be called the subgraph induced in $G$ by $S$, or (if context makes the choice of $G$ unambiguous) the induced subgraph of $S$.

## Examples

Important types of induced subgraphs include the following:

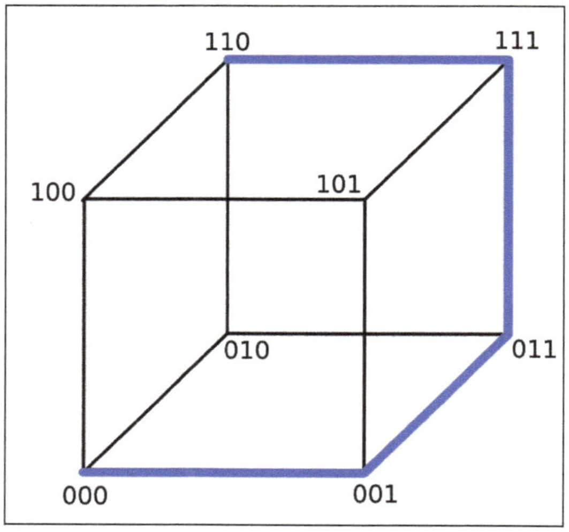


The snake-in-the-box problem concerns the longest induced paths in hypercube graphs.

- Induced paths are induced subgraphs that are paths. The shortest path between any two vertices in an unweighted graph is always an induced path, because any additional edges between pairs of vertices that could cause it to be not induced would also cause it to be not shortest. Conversely, in distance-hereditary graphs, every induced path is a shortest path.
- Induced cycles are induced subgraphs or cycles. The girth of a graph is defined by the length of its shortest cycle, which is always an induced cycle. According to the strong perfect graph theorem, induced cycles and their complements play a critical role in the characterization of perfect graphs.
- Cliques and independent sets are induced subgraphs that are respectively complete graphs or edgeless graphs.
- Induced matchings are induced subgraphs that are matchings.
- The neighborhood of a vertex is the induced subgraph of all vertices adjacent to it.

## Computation

The induced subgraph isomorphism problem is a form of the subgraph isomorphism problem in which the goal is to test whether one graph can be found as an induced subgraph of another. Because it includes the clique problem as a special case, it is NP-complete.

# Complete Graph

In the mathematical field of graph theory, a complete graph is a simple undirected graph in which every pair of distinct vertices is connected by a unique edge. A complete digraph is a directed graph in which every pair of distinct vertices is connected by a pair of unique edges (one in each direction).

Graph theory itself is typically dated as beginning with Leonhard Euler's 1736 work on the Seven Bridges of Königsberg. However, drawings of complete graphs, with their vertices placed on the points of a regular polygon, appeared already in the 13th century, in the work of Ramon Llull. Such a drawing is sometimes referred to as a mystic rose.

## Properties

The complete graph on $n$ vertices is denoted by $K_n$. Some sources claim that the letter K in this notation stands for the German word *komplett*, but the German name for a complete graph, *vollständiger Graph*, does not contain the letter K, and other sources state that the notation honors the contributions of Kazimierz Kuratowski to graph theory.

$K_n$ has $n(n - 1)/2$ edges (a triangular number), and is a regular graph of degree $n - 1$. All complete graphs are their own maximal cliques. They are maximally connected as the only vertex cut which disconnects the graph is the complete set of vertices. The complement graph of a complete graph is an empty graph.

If the edges of a complete graph are each given an orientation, the resulting directed graph is called a tournament.

The number of matchings of the complete graphs are given by the telephone numbers:

> 1, 1, 2, 4, 10, 26, 76, 232, 764, 2620, 9496, 35696, 140152, 568504, 2390480, 10349536, 46206736, ... (sequence A000085 in the OEIS).

These numbers give the largest possible value of the Hosoya index for an $n$-vertex graph. The number of perfect matchings of the complete graph $K_n$ (with $n$ even) is given by the double factorial $(n - 1)!!$.

The crossing numbers up to $K_{27}$ are known, with $K_{28}$ requiring either 7233 or 7234 crossings. Further values are collected by the Rectilinear Crossing Number project. Rectilinear Crossing numbers for $K_n$ are:

> 0, 0, 0, 0, 1, 3, 9, 19, 36, 62, 102, 153, 229, 324, 447, 603, 798, 1029, 1318, 1657, 2055, 2528, 3077, 3699, 4430, 5250, 6180, ... (sequence A014540 in the OEIS).

## Geometry and Topology

A complete graph with $n$ nodes represents the edges of an $(n - 1)$-simplex. Geometrically $K_3$ forms the edge set of a triangle, $K_4$ a tetrahedron, etc. The Császár polyhedron, a nonconvex polyhedron with the topology of a torus, has the complete graph $K_7$ as its skeleton. Every neighborly polytope in four or more dimensions also has a complete skeleton.

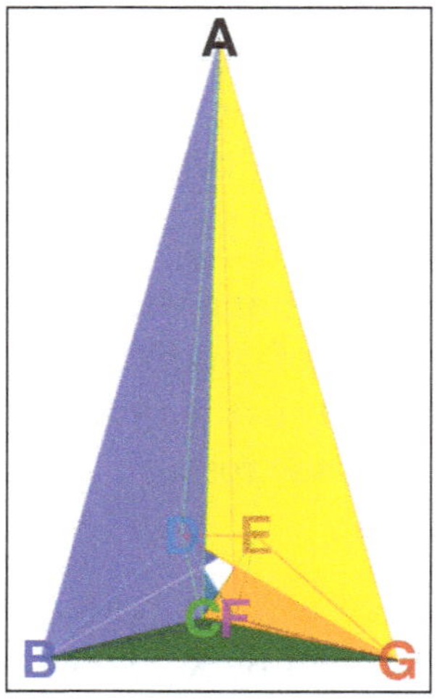


Interactive Csaszar polyhedron model with vertices representing nodes.
In the SVG image, move the mouse to rotate it.

$K_1$ through $K_4$ are all planar graphs. However, every planar drawing of a complete graph with five or more vertices must contain a crossing, and the nonplanar complete graph $K_5$ plays a key role in the characterizations of planar graphs: by Kuratowski's theorem, a graph is planar if and only if it contains neither $K_5$ nor the complete bipartite graph $K_{3,3}$ as a subdivision, and by Wagner's theorem the same result holds for graph minors in place of subdivisions. As part of the Petersen family, $K_6$ plays a similar role as one of the forbidden minors for linkless embedding. In other words, and as Conway and Gordon proved, every embedding of $K_6$ into three-dimensional space is intrinsically linked, with at least one pair of linked triangles. Conway and Gordon also showed that any three-dimensional embedding of $K_7$ contains a Hamiltonian cycle that is embedded in space as a nontrivial knot.

## Examples

Complete graphs on $n$ vertices, for $n$ between 1 and 12, are shown below along with the numbers of edges:

| $K_1$: 0 | $K_2$: 1 | $K_3$: 3 | $K_4$: 6 |
|---|---|---|---|
|  |  | 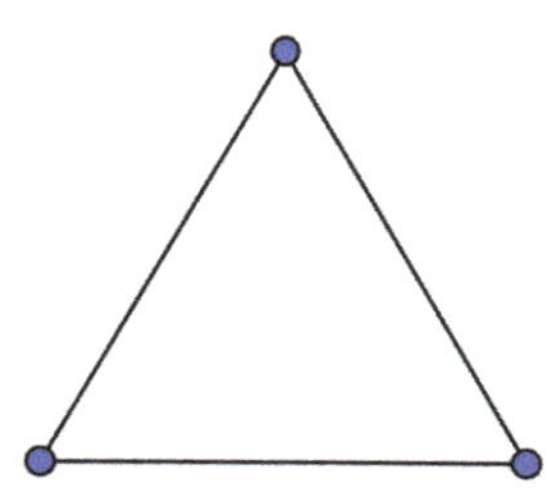 | 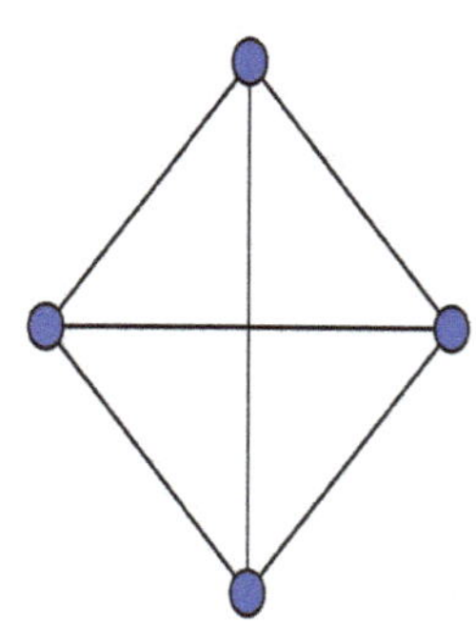 |

| $K_5$: 10 | $K_6$: 15 | $K_7$: 21 | $K_8$: 28 |
|---|---|---|---|
| 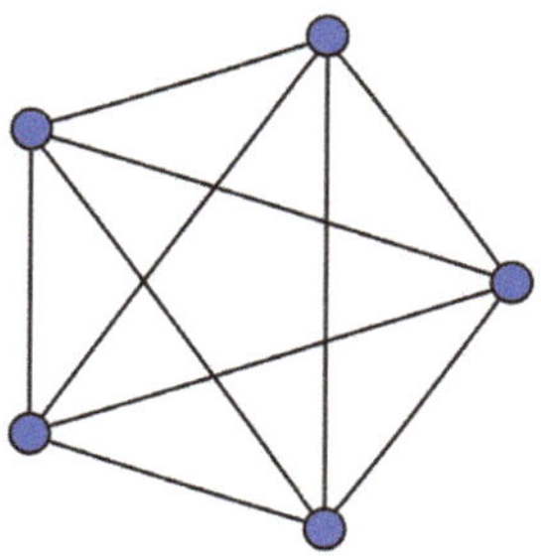 | 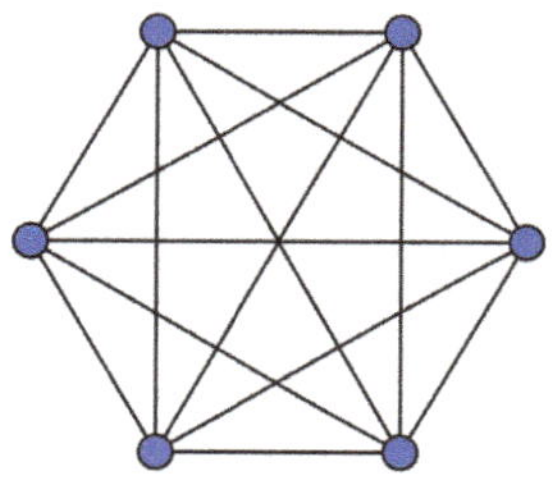 | 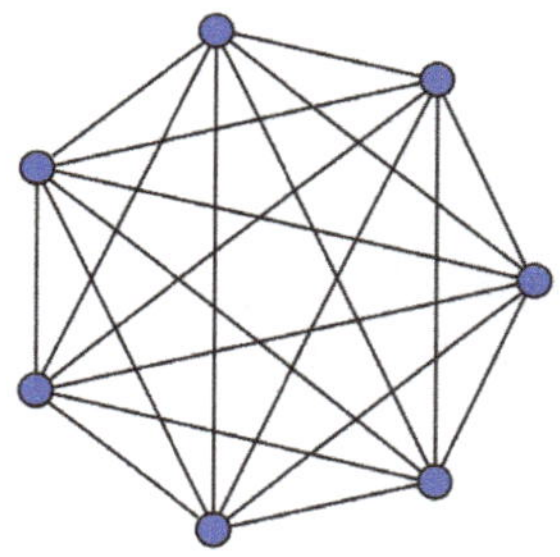 | 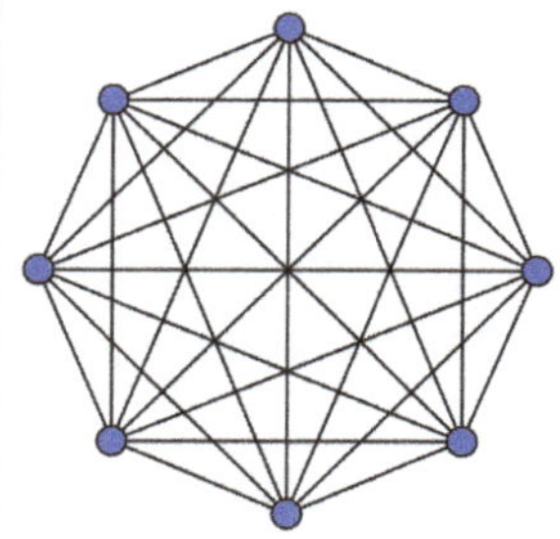 |
| $K_9$: 36 | $K_{10}$: 45 | $K_{11}$: 55 | $K_{12}$: 66 |
| 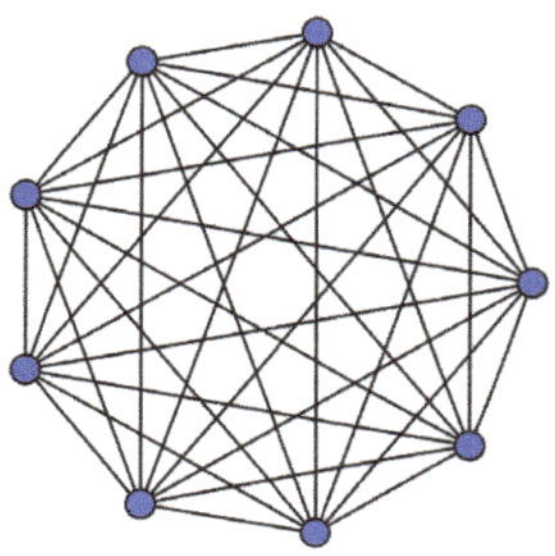 | 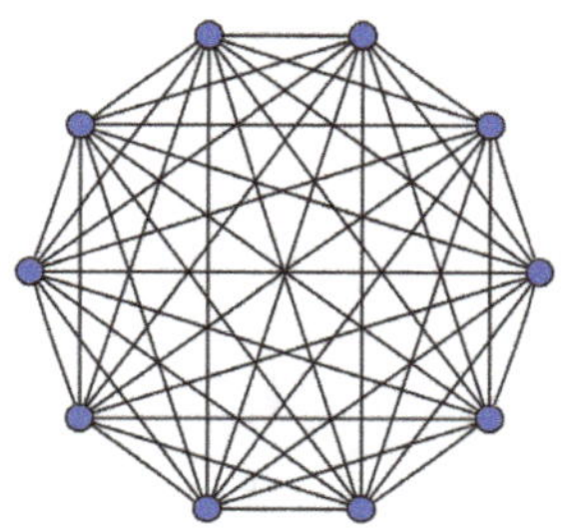 | 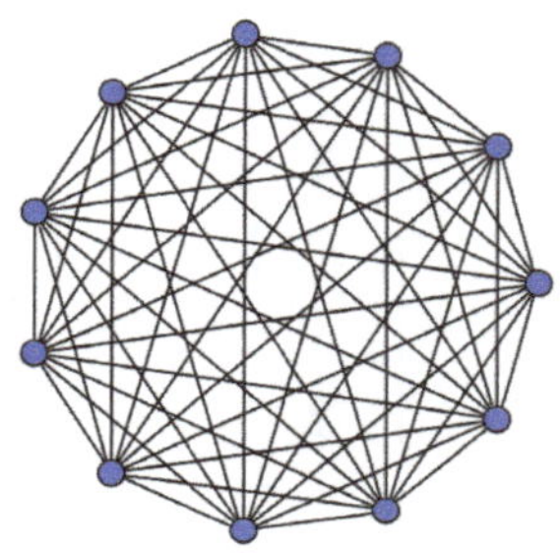 | 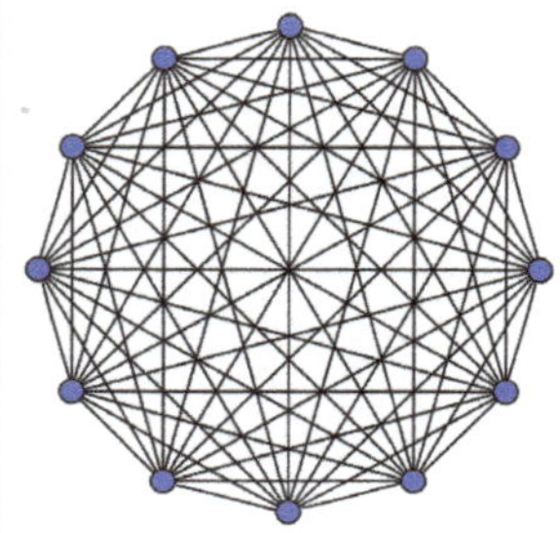 |

## Directed Graph

In mathematics, and more specifically in graph theory, a directed graph (or digraph) is a graph that is made up of a set of vertices connected by edges, where the edges have a direction associated with them.

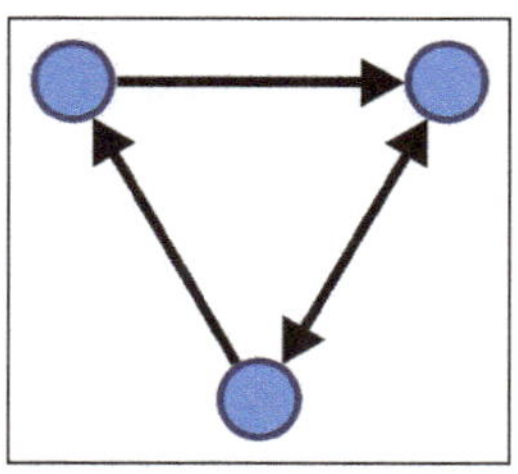

A simple directed graph.

In formal terms, a directed graph is an ordered pair $G = (V, A)$ where:

- $V$ is a set whose elements are called *vertices*, *nodes*, or *points*;
- $A$ is a set of ordered pairs of vertices, called *arrows*, *directed edges* (sometimes simply *edges* with the corresponding set named $E$ instead of $A$), *directed arcs*, or *directed lines*.

It differs from an ordinary or undirected graph, in that the latter is defined in terms of unordered pairs of vertices, which are usually called *edges*, *arcs*, or *lines*.

The aforementioned definition does not allow a directed graph to have multiple arrows with the same source and target nodes, but some authors consider a broader definition that allows directed graphs to have such multiple arrows (namely, they allow the arrows set to be a multiset). More specifically, these entities are addressed as directed multigraphs (or multidigraphs). On the other hand, the aforementioned definition allows a directed graph to have loops (that is, arrows that connect nodes with themselves), but some authors consider a narrower definition that doesn't allow directed graphs to have loops. More specifically, directed graphs without loops are addressed as simple directed graphs, while directed graphs with loops are addressed as *loop-digraphs*.

## Types of Directed Graphs

### Subclasses

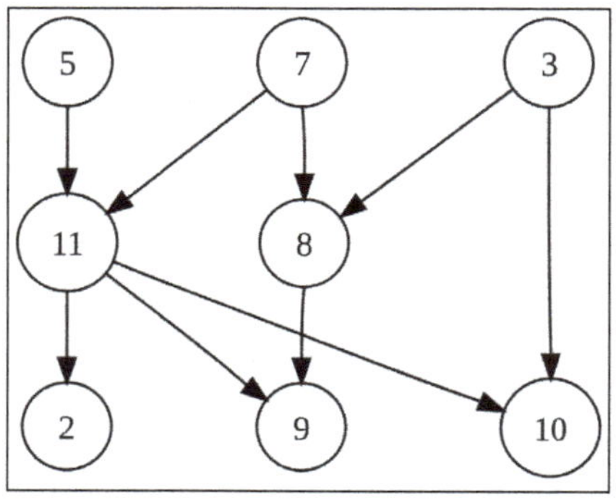


A simple directed acyclic graph.

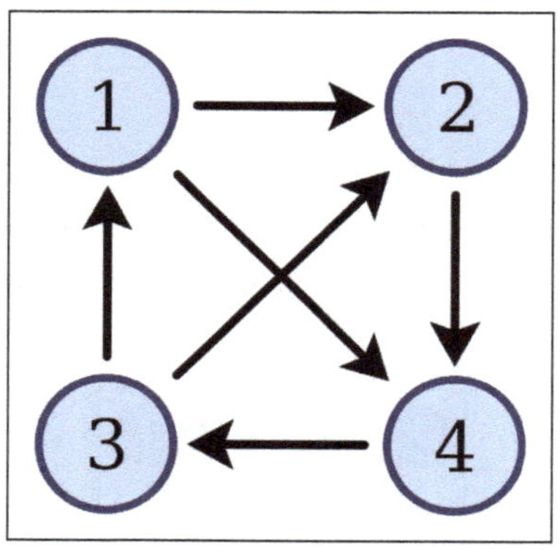


A tournament on 4 vertices.

- Symmetric directed graphs are directed graphs where all edges are bidirected (that is, for every arrow that belongs to the digraph, the corresponding inversed arrow also belongs to it).
- Simple directed graphs are directed graphs that have no loops (arrows that connect vertices to themselves) and no multiple arrows with same source and target nodes. As already introduced, in case of multiple arrows the entity is usually addressed as *directed multigraph*. Some authors describe digraphs with loops as *loop-digraphs*.
    - Complete directed graphs are simple directed graphs where each pair of vertices is joined by a symmetric pair of directed arrows (it is equivalent to an undirected complete

graph with the edges replaced by pairs of inverse arrows). It follows that a complete digraph is symmetric.

- Oriented graphs are directed graphs having no bidirected edges (i.e. at most one of $(x, y)$ and $(y, x)$ may be arrows of the graph). It follows that a directed graph is an oriented graph if and only if it hasn't any 2-cycle.
    - Tournaments are oriented graphs obtained by choosing a direction for each edge in undirected complete graphs.
    - Directed acyclic graphs (DAGs) are directed graphs with no directed cycles.
        - *Multitrees* are DAGs in which no two directed paths from a single starting vertex meet back at the same ending vertex.
        - *Oriented trees* or *polytrees* are DAGs formed by orienting the edges of undirected acyclic graphs.
            - *Rooted trees* are oriented trees in which all edges of the underlying undirected tree are directed either away from or towards the root.

## Digraphs with Supplementary Properties

- Weighted directed graphs (also known as directed networks) are (simple) directed graphs with *weights* assigned to their arrows, similarly to weighted graphs (which are also known as undirected networks or weighted networks).
    - Flow networks are weighted directed graphs where two nodes are distinguished, a *source* and a *sink*.
- Rooted directed graphs (also known as flow graphs) are digraphs in which a vertex has been distinguished as the root.
    - *Control flow graphs* are rooted digraphs used in computer science as a representation of the paths that might be traversed through a program during its execution.
- Signal-flow graphs are directed graphs in which nodes represent system variables and branches (edges, arcs, or arrows) represent functional connections between pairs of nodes.
- Flow graphs are digraphs associated with a set of linear algebraic or differential equations.
- State diagrams are directed multigraphs that represent finite state machines.
- Commutative diagrams are digraphs used in category theory, where the vertices represent (mathematical) objects and the arrows represent morphisms, with the property that all directed paths with the same start and endpoints lead to the same result by composition.
- In the theory of Lie groups, a quiver $Q$ is a directed graph serving as the domain of, and thus characterizing the shape of, a *representation* $V$ defined as a functor, specifically an object of the functor category $\mathrm{FinVct}_K^{F(Q)}$ where $F(Q)$ is the free category on $Q$ consisting

of paths in $Q$ and $\mathrm{FinVct}_K$ is the category of finite-dimensional vector spaces over a field $K$. Representations of a quiver label its vertices with vector spaces and its edges (and hence paths) compatibly with linear transformations between them, and transform via natural transformations.

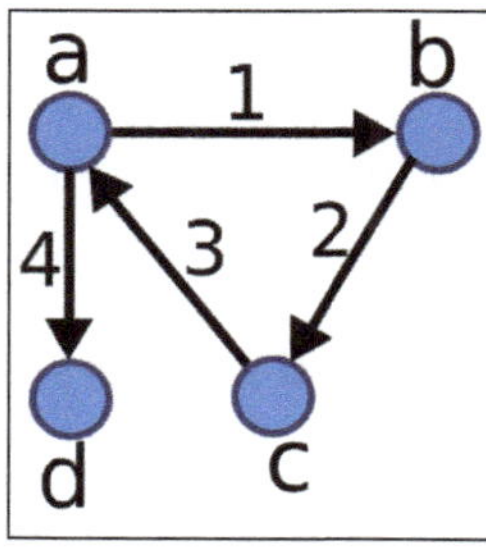


| | 1 | 2 | 3 | 4 |
|---|---|---|---|---|
| a | 1 | 0 | -1 | 1 |
| b | -1 | 1 | 0 | 0 |
| c | 0 | -1 | 1 | 0 |
| d | 0 | 0 | 0 | -1 |

Oriented graph with corresponding incidence matrix.

An arrow $(x, y)$ is considered to be directed *from $x$ to $y$*; $y$ is called the *head* and $x$ is called the *tail* of the arrow; $y$ is said to be a *direct successor* of $x$ and $x$ is said to be a *direct predecessor* of $y$. If a path leads from $x$ to $y$, then $y$ is said to be a *successor* of $x$ and *reachable* from $x$, and $x$ is said to be a *predecessor* of $y$. The arrow $(y, x)$ is called the *inverted arrow* of $(x, y)$.

The adjacency matrix of a multidigraph with loops is the integer-valued matrix with rows and columns corresponding to the vertices, where a nondiagonal entry $a_{ij}$ is the number of arrows from vertex $i$ to vertex $j$, and the diagonal entry $a_{ii}$ is the number of loops at vertex $i$. The adjacency matrix of a directed graph is unique up to identical permutation of rows and columns.

Another matrix representation for a directed graph is its incidence matrix.

## Indegree and Outdegree

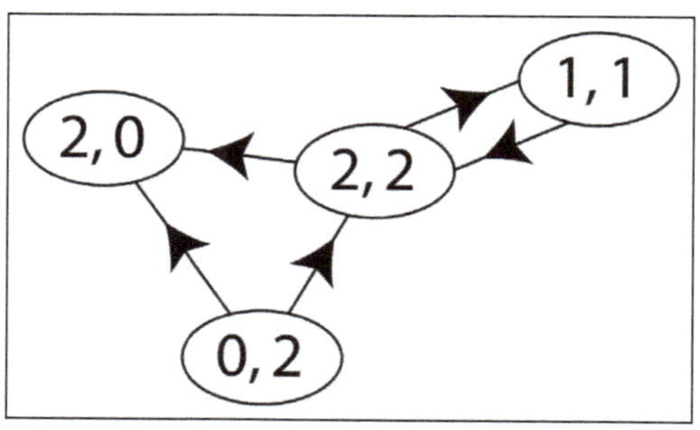


A directed graph with vertices labeled (indegree, outdegree).

For a vertex, the number of head ends adjacent to a vertex is called the *indegree* of the vertex and the number of tail ends adjacent to a vertex is its *outdegree* (called "branching factor" in trees).

Let $G = (V, A)$ and $v \in V$. The indegree of $v$ is denoted $\deg^-(v)$ and its outdegree is denoted $\deg^+(v)$.

A vertex with $\deg^-(v) = 0$ is called a *source*, as it is the origin of each of its outcoming arrows. Similarly, a vertex with $\deg^+(v) = 0$ is called a *sink*, since it is the end of each of its incoming arrows.

If a vertex is neither a *source* nor a *sink*, it is called an *internal*.

The *degree sum formula* states that, for a directed graph,

$$\sum_{v \in V} \deg^-(v) = \sum_{v \in V} \deg^+(v) = |A|.$$

If for every vertex $v \in V$, $\deg^+(v) = \deg^-(v)$, the graph is called a *balanced directed graph*.

### Degree Sequence

The degree sequence of a directed graph is the list of its indegree and outdegree pairs; for the above example we have degree sequence ((2, 0), (2, 2), (0, 2), (1, 1)). The degree sequence is a directed graph invariant so isomorphic directed graphs have the same degree sequence. However, the degree sequence does not, in general, uniquely identify a directed graph; in some cases, non-isomorphic digraphs have the same degree sequence.

The directed graph realization problem is the problem of finding a directed graph with the degree sequence a given sequence of positive integer pairs. (Trailing pairs of zeros may be ignored since they are trivially realized by adding an appropriate number of isolated vertices to the directed graph.) A sequence which is the degree sequence of some directed graph, i.e. for which the directed graph realization problem has a solution, is called a directed graphic or directed graphical sequence. This problem can either be solved by the Kleitman–Wang algorithm or by the Fulkerson–Chen–Anstee theorem.

### Directed Graph Connectivity

A directed graph is *weakly connected* (or just *connected*) if the undirected *underlying graph* obtained by replacing all directed edges of the graph with undirected edges is a connected graph. A directed graph is *strongly connected* or *strong* if it contains a directed path from $x$ to $y$ and a directed path from $y$ to $x$ for every pair of vertices $\{x, y\}$. The *strong components* are the maximal strongly connected subgraphs.

## Undirected Graph

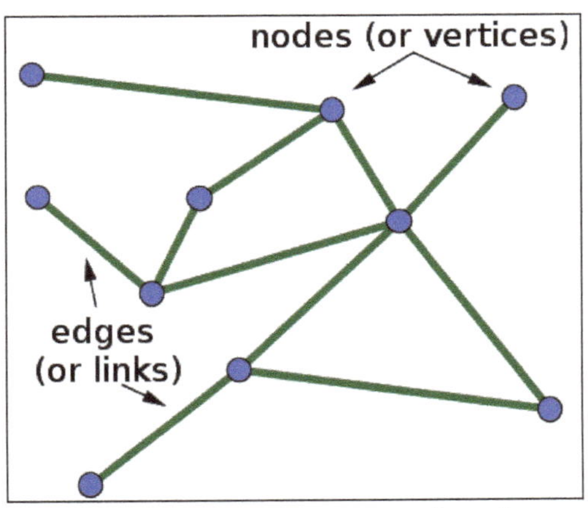


An undirected graph with 10 and 11 edges.

An undirected graph is graph, i.e., a set of objects (called vertices or nodes) that are connected together, where all the edges are bidirectional. An undirected graph is sometimes called an undirected network. In contrast, a graph where the edges point in a direction is called a directed graph.

One can formally define an undirected graph as $G=(\mathcal{N},\mathcal{E})$, consisting of the set $\mathcal{N}$ of nodes and the set EE of edges, which are unordered pairs of elements of $\mathcal{N}$.

## References


- Khot, Subhash; Regev, Oded (2008). "Vertex cover might be hard to approximate to within 2−ε". Journal of Computer and System Sciences. 74 (3): 335–349. Doi:10.1016/j.jcss.2007.06.019
- Mathematics-graph-theory-basics: geeksforgeeks.org, Retrieved 5 February, 2019
- Dinur, Irit; Safra, Samuel (2005). "On the hardness of approximating minimum vertex cover". Annals of Mathematics. 162 (1): 439–485. Citeseerx 10.1.1.125.334. Doi:10.4007/annals.2005.162.439
- Bonato, Anthony (2008), A course on the web graph, Graduate Studies in Mathematics, 89, Atlantic Association for Research in the Mathematical Sciences (AARMS), Halifax, NS, p. 7, doi:10.1090/gsm/089, ISBN 978-0-8218-4467-0, MR 2389013
- Curtin, Brian (2005), "Algebraic characterizations of graph regularity conditions", Designs, Codes and Cryptography, 34 (2–3): 241–248, doi:10.1007/s10623-004-4857-4, MR 2128333
- Nešetřil, Jaroslav; Ossona de Mendez, Patrice (2012), "3.10 Graph Parameters", Sparsity: Graphs, Structures, and Algorithms, Algorithms and Combinatorics, 28, Springer, pp. 54–56, doi:10.1007/978-3-642-27875-4, ISBN 978-3-642-27874-7, MR 2920058
- Grohe, Martin (2007), "The complexity of homomorphism and constraint satisfaction problems seen from the other side", J. ACM, 54 (1), doi:10.1145/1206035.1206036

# 6

# Probability

The branch of mathematics that is concerned with the study of possible outcomes of dependent and independent events is known as probability. It includes probability distribution, binomial distribution, Bayes' theorem, conditional probability etc. This chapter has been carefully written to provide an easy understanding of these aspects of probability.

Probability is the likelihood of something happening. When someone tells you the probability of something happening, they are telling you how likely that something is. When people buy lottery tickets, the probability of winning is usually stated, and sometimes, it can be something like 1/10,000,000 (or even worse). This tells you that it is not very likely that you will win.

The formula for probability tells you how many choices you have over the number of possible combinations.

$$probability = \frac{possible\,choices}{total\,number\,of\,options}$$

## Calculating Probability

To calculate probability, you need to know how many possible options or outcomes there are and how many right combinations you have. Let's calculate the probability of throwing dice to see how it works.

First, we know that a die has a total of 6 possible outcomes. You can roll a 1, 2, 3, 4, 5, or 6. Next, we need to know how many choices we have. Whenever you roll, you will get one of the numbers. You can't roll and get two different numbers with one die. So, our number of choices is 1. Using formula for probability, we get a probability of 1/6.

$$probability = \frac{1}{6}$$

Our probability of rolling any of the numbers is 1/6. The probability of rolling a 2 is 1/6, of rolling a 3 is also 1/6, and so on.

Let's try another problem. Let's say we have a grab bag of apples and oranges. We want to find out the probability of picking an apple from the bag. One thing we need to know is the number of apples in the bag because that gives us the number of 'correct' choices, which is the number of our possible choices in the top part of the calculation.

We also need to know the total number of fruits in the bag, for this gives us the total number of choices we have, or the total number of options in the bottom part of the calculation. The person with the grab bag tells us there are 10 apples and 20 oranges in the bag. So, what is our probability of picking an apple? We have 10 apples, one of which we want, and a total of 30 fruits to pick from.

$$\begin{aligned} probability &= \frac{10}{6} \\ &= \frac{1}{3} \end{aligned}$$

Our probability is 1/3 for picking an apple. If you compare this with our probability of rolling a number on a die, the probability of picking an apple from the grab bag is higher. It is more likely that we will pick an apple than that we will roll a particular number.

In both cases, we can leave the probability in fraction form or we can convert it to decimal form: 1/6 becomes 0.17, and 1/3 becomes 0.33.

Example: What is the probability to get a 6 when you roll a die?

A die has 6 sides, 1 side contain the number 6 that give us 1 wanted outcome in 6 possible outcomes.

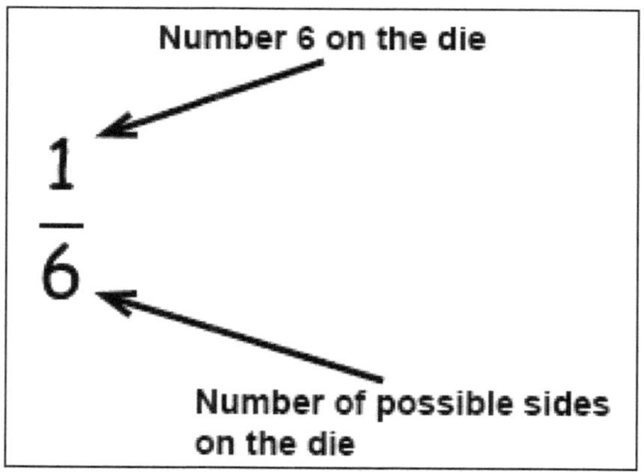


Independent events: Two events are independent when the outcome of the first event does not influence the outcome of the second event.

When we determine the probability of two independent events we multiply the probability of the first event by the probability of the second event.

$$P(X \, and \, Y) = P(X).P(Y)$$

To find the probability of an independent event we are using this rule:

Example: If one has three dice what is the probability of getting three 4s?

The probability of getting a 4 on one die is 1/6.

The probability of getting 3 4s is:

$$P(4\,and\,4\,and\,4) = \frac{1}{6}.\frac{1}{6}.\frac{1}{6} = \frac{1}{216}$$

When the outcome affects the second outcome, which is what we called dependent events.

Dependent events: Two events are dependent when the outcome of the first event influences the outcome of the second event. The probability of two dependent events is the product of the probability of X and the probability of Y after X occurs.

$$P(X\,and\,Y) = P(X).P(Y\,after\,x)$$

Example: What is the probability for you to choose two red cards in a deck of cards?

A deck of cards has 26 black and 26 red cards. The probability of choosing a red card randomly is:

$$P(red) = \frac{26}{52} = \frac{1}{2}$$

The probability of choosing a second red card from the deck is now:

$$P(red) = \frac{25}{51}$$

The probability:

$$P(2red) = \frac{1}{2}.\frac{25}{51} = \frac{25}{102}$$

Two events are mutually exclusive when two events cannot happen at the same time. The probability that one of the mutually exclusive events occur is the sum of their individual probabilities.

$$P(X\,or\,Y) = P(X) + P(Y)$$

An example of two mutually exclusive events is a wheel of fortune. Let's say you win a bar of chocolate if you end up in a red or a pink field.

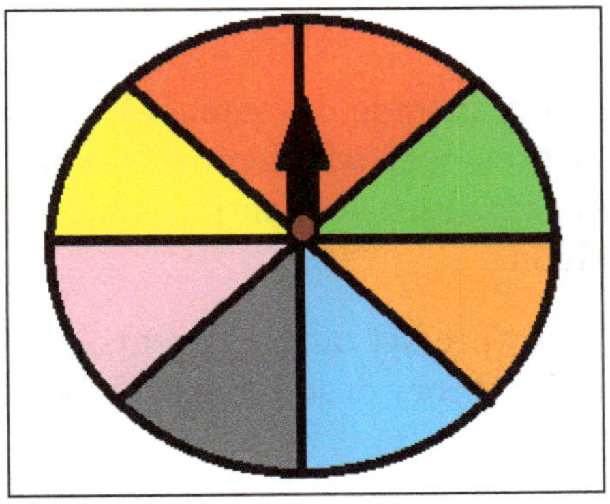

What is the probability that the wheel stops at red or pink?

P(red or pink)=P(red)+P(pink)

$$P(red)=\frac{2}{8}=\frac{1}{4}$$

$$P(pink)=\frac{1}{8}$$

$$P(red\,or\,pink)=\frac{1}{8}+\frac{2}{8}=\frac{3}{8}$$

Inclusive events are events that can happen at the same time. To find the probability of an inclusive event we first add the probabilities of the individual events and then subtract the probability of the two events happening at the same time.

$$P(X\,or\,Y)=P(X)+P(Y)-P(X\,and\,Y)$$

Example: What is the probability of drawing a black card or a ten in a deck of cards?

There are 4 tens in a deck of cards P(10) = 4/52.

There are 26 black cards P(black) = 26/52.

There are 2 black tens P(black and 10) = 2/52.

$$p(black\,or\,ten)=\frac{4}{52}+\frac{26}{52}-\frac{2}{52}=\frac{30}{52}-\frac{2}{52}=\frac{28}{52}=\frac{7}{13}$$

## Probability Theory

Probability theory is the branch of mathematics concerned with probability. Although there are several different probability interpretations, probability theory treats the concept in a rigorous mathematical manner by expressing it through a set of axioms. Typically these axioms formalise probability in terms of a probability space, which assigns a measure taking values between 0 and 1, termed the probability measure, to a set of outcomes called the sample space. Any specified subset of these outcomes is called an event.

Central subjects in probability theory include discrete and continuous random variables, probability distributions, and stochastic processes, which provide mathematical abstractions of non-deterministic or uncertain processes or measured quantities that may either be single occurrences or evolve over time in a random fashion.

Although it is not possible to perfectly predict random events, much can be said about their behavior. Two major results in probability theory describing such behaviour are the law of large numbers and the central limit theorem.

As a mathematical foundation for statistics, probability theory is essential to many human activities that involve quantitative analysis of data. Methods of probability theory also apply to descriptions of complex systems given only partial knowledge of their state, as in statistical mechanics. A great discovery of twentieth-century physics was the probabilistic nature of physical phenomena at atomic scales.

## Treatment

Most introductions to probability theory treat discrete probability distributions and continuous probability distributions separately. The measure theory-based treatment of probability covers the discrete, continuous, a mix of the two, and more.

Consider an experiment that can produce a number of outcomes. The set of all outcomes is called the *sample space* of the experiment. The *power set* of the sample space (or equivalently, the event space) is formed by considering all different collections of possible results. For example, rolling an honest die produces one of six possible results. One collection of possible results corresponds to getting an odd number. Thus, the subset {1,3,5} is an element of the power set of the sample space of die rolls. These collections are called *events*. In this case, {1,3,5} is the event that the die falls on some odd number. If the results that actually occur fall in a given event, that event is said to have occurred.

Probability is a way of assigning every "event" a value between zero and one, with the requirement that the event made up of all possible results (in example, the event {1,2,3,4,5,6}) be assigned a value of one. To qualify as a probability distribution, the assignment of values must satisfy the requirement that if you look at a collection of mutually exclusive events (events that contain no common results, e.g., the events {1,6}, {3}, and {2,4} are all mutually exclusive), the probability that any of these events occurs is given by the sum of the probabilities of the events.

The probability that any one of the events {1,6}, {3}, or {2,4} will occur is 5/6. This is the same as saying that the probability of event {1,2,3,4,6} is 5/6. This event encompasses the possibility of any number except five being rolled. The mutually exclusive event {5} has a probability of 1/6, and the event {1,2,3,4,5,6} has a probability of 1, that is, absolute certainty.

When doing calculations using the outcomes of an experiment, it is necessary that all those elementary events have a number assigned to them. This is done using a random variable. A random variable is a function that assigns to each elementary event in the sample space a real number. This function is usually denoted by a capital letter. In the case of a die, the assignment of a number to a certain elementary events can be done using the identity function. This does not always work. For example, when flipping a coin the two possible outcomes are "heads" and "tails". In this example, the random variable *X* could assign to the outcome "heads" the number "0" ( $X(heads) = 0$ ) and to the outcome "tails" the number "1" ( $X(tails) = 1$ ).

## Discrete Probability Distributions

*Discrete probability theory* deals with events that occur in countable sample spaces.

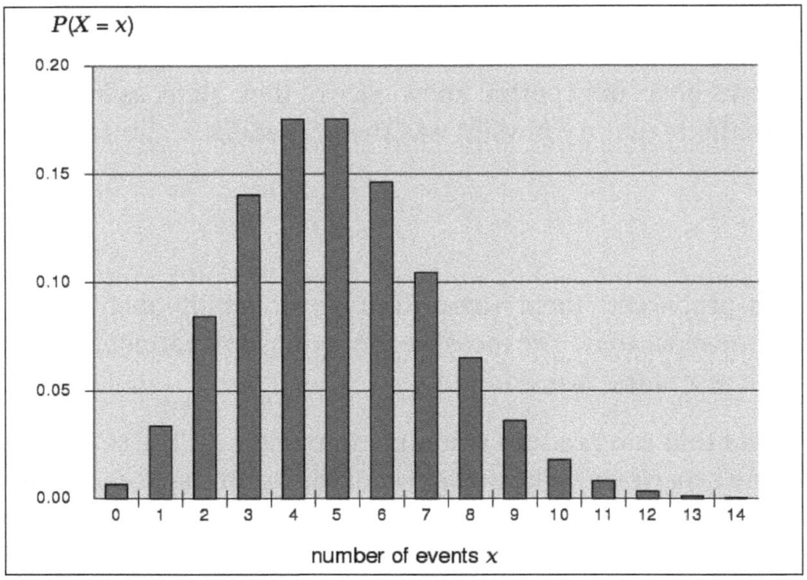


The Poisson distribution, a discrete probability distribution.

Examples: Throwing dice, experiments with decks of cards, random walk, and tossing coins.

*Classical definition*: Initially the probability of an event to occur was defined as the number of cases favorable for the event, over the number of total outcomes possible in an equiprobable sample space.

For example, if the event is "occurrence of an even number when a die is rolled", the probability is given by $\frac{3}{6}=\frac{1}{2}$, since 3 faces out of the 6 have even numbers and each face has the same probability of appearing.

*Modern definition*: The modern definition starts with a finite or countable set called the sample space, which relates to the set of all *possible outcomes* in classical sense, denoted by $\Omega$. It is then assumed that for each element $x \in \Omega$, an intrinsic "probability" value $f(x)$ is attached, which satisfies the following properties:

1. $f(x) \in [0,1]$ for all $x \in \Omega$;
2. $\sum_{x \in \Omega} f(x) = 1.$

That is, the probability function $f(x)$ lies between zero and one for every value of $x$ in the sample space $\Omega$, and the sum of $f(x)$ over all values $x$ in the sample space $\Omega$ is equal to 1. An *event* is defined as any subset $E$ of the sample space $\Omega$. The *probability* of the event $E$ is defined as:

$$P(E) = \sum_{x \in E} f(x)$$

So, the probability of the entire sample space is 1, and the probability of the null event is 0.

The function $f(x)$ mapping a point in the sample space to the "probability" value is called a *probability mass function* abbreviated as *pmf*. The modern definition does not try to answer how probability mass functions are obtained; instead, it builds a theory that assumes their existence.

## Continuous Probability Distributions

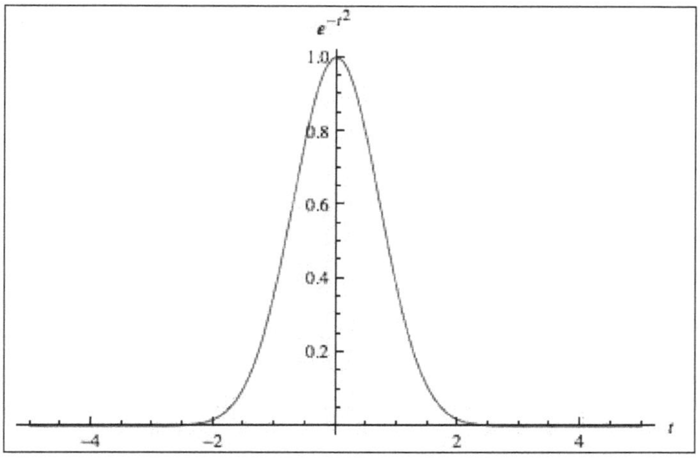


The normal distribution, a continuous probability distribution.

*Continuous probability theory* deals with events that occur in a continuous sample space.

*Classical definition*: The classical definition breaks down when confronted with the continuous case.

*Modern definition*: If the outcome space of a random variable $X$ is the set of real numbers ($\mathbb{R}$) or a subset thereof, then a function called the *cumulative distribution function* (or *cdf*) $F$ exists, defined by $F(x) = P(X \leq x)$. That is, $F(x)$ returns the probability that $X$ will be less than or equal to $x$.

The cdf necessarily satisfies the following properties.

1. $F$ is a monotonically non-decreasing, right-continuous function;
2. $\lim_{x \to -\infty} F(x) = 0;$
3. $\lim_{x \to \infty} F(x) = 1.$

If $F$ is absolutely continuous, i.e., its derivative exists and integrating the derivative gives us the cdf back again, then the random variable $X$ is said to have a *probability density function* or *pdf* or simply *density* $f(x) = \frac{dF(x)}{dx}$.

For a set $E \subseteq \mathbb{R}$, the probability of the random variable $X$ being in $E$ is:

$$P(X \in E) = \int_{x \in E} dF(x).$$

In case the probability density function exists, this can be written as:

$$P(X \in E) = \int_{x \in E} f(x)dx.$$

Whereas the *pdf* exists only for continuous random variables, the *cdf* exists for all random variables (including discrete random variables) that take values in $\mathbb{R}$.

These concepts can be generalized for multidimensional cases on $\mathbb{R}^n$ and other continuous sample spaces.

## Measure-theoretic Probability Theory

The *raison d'être* of the measure-theoretic treatment of probability is that it unifies the discrete and the continuous cases, and makes the difference a question of which measure is used. Furthermore, it covers distributions that are neither discrete nor continuous nor mixtures of the two.

An example of such distributions could be a mix of discrete and continuous distributions—for example, a random variable that is 0 with probability 1/2, and takes a random value from a normal distribution with probability 1/2. It can still be studied to some extent by considering it to have a pdf of $(\delta[x]+\varphi(x))/2$, where $\delta[x]$ is the Dirac delta function.

Other distributions may not even be a mix, for example, the Cantor distribution has no positive probability for any single point, neither does it have a density. The modern approach to probability theory solves these problems using measure theory to define the probability space:

Given any set $\Omega$ (also called *sample space*) and a σ-algebra $\mathcal{F}$ on it, a measure $P$ defined on $\mathcal{F}$ is called a *probability measure* if $P(\Omega)=1$.

If $\mathcal{F}$ is the Borel σ-algebra on the set of real numbers, then there is a unique probability measure on $\mathcal{F}$ for any cdf, and vice versa. The measure corresponding to a cdf is said to be *induced* by the cdf. This measure coincides with the pmf for discrete variables and pdf for continuous variables, making the measure-theoretic approach free of fallacies.

The *probability* of a set $E$ in the σ-algebra $\mathcal{F}$ is defined as,

$$P(E)=\int_{\omega\in E}\mu_F(d\omega)$$

where the integration is with respect to the measure $\mu_F$ induced by $F$.

Along with providing better understanding and unification of discrete and continuous probabilities, measure-theoretic treatment also allows us to work on probabilities outside $\mathbb{R}^n$, as in the theory of stochastic processes. For example, to study Brownian motion, probability is defined on a space of functions.

When it's convenient to work with a dominating measure, the Radon-Nikodym theorem is used to define a density as the Radon-Nikodym derivative of the probability distribution of interest with respect to this dominating measure. Discrete densities are usually defined as this derivative with respect to a counting measure over the set of all possible outcomes. Densities for absolutely continuous distributions are usually defined as this derivative with respect to the Lebesgue measure. If a theorem can be proved in this general setting, it holds for both discrete and continuous distributions as well as others; separate proofs are not required for discrete and continuous distributions.

## Classical Probability Distributions

Certain random variables occur very often in probability theory because they well describe many natural or physical processes. Their distributions, therefore, have gained *special importance* in probability theory. Some fundamental *discrete distributions* are the discrete uniform, Bernoulli, binomial, negative binomial, Poisson and geometric distributions. Important *continuous distributions* include the continuous uniform, normal, exponential, gamma and beta distributions.

## Convergence of Random Variables

In probability theory, there are several notions of convergence for random variables. They are listed below in the order of strength, i.e., any subsequent notion of convergence in the list implies convergence according to all of the preceding notions.

### Weak Convergence

A sequence of random variables $X_1, X_2, \ldots,$ converges *weakly* to the random variable $X$ if their respective cumulative *distribution functions* $F_1, F_2, \ldots$ converge to the cumulative distribution function $F$ of $X$, wherever $F$ is continuous. Weak convergence is also called *convergence in distribution.*

Most common shorthand notation: $X_n \xrightarrow{\mathcal{D}} X$.

### Convergence in Probability

The sequence of random variables $X_1, X_2, \ldots$ is said to converge towards the random variable $X$ *in probability* if $\lim_{n\to\infty} P\left(\left|X_n - X\right| \geq \varepsilon\right) = 0$ for every ε > 0.

Most common shorthand notation: $X_n \xrightarrow{P} X$.

### Strong Convergence

The sequence of random variables $X_1, X_2, \ldots$ is said to converge towards the random variable $X$ *strongly* if $P(\lim_{n\to\infty} X_n = X) = 1$. Strong convergence is also known as *almost sure convergence.*

Most common shorthand notation: $X_n \xrightarrow{\text{a.s.}} X$.

As the names indicate, weak convergence is weaker than strong convergence. In fact, strong convergence implies convergence in probability, and convergence in probability implies weak convergence. The reverse statements are not always true.

## Law of Large Numbers

Common intuition suggests that if a fair coin is tossed many times, then *roughly* half of the time it will turn up *heads*, and the other half it will turn up *tails*. Furthermore, the more often the coin is tossed, the more likely it should be that the ratio of the number of *heads* to the number of *tails* will approach unity. Modern probability theory provides a formal version of this intuitive idea, known as the *law of large numbers*. This law is remarkable because it is not assumed in the foundations of probability theory, but instead emerges from these foundations as a theorem. Since it links theoretically derived probabilities to their actual frequency of occurrence in the real world, the law of large numbers is considered as a pillar in the history of statistical theory and has had widespread influence.

The *law of large numbers* (LLN) states that the sample average,

$$\overline{X}_n = \frac{1}{n}\sum_{k=1}^{n} X_k$$

of a sequence of independent and identically distributed random variables $X_k$ converges towards their common expectation $\mu$, provided that the expectation of $|X_k|$ is finite.

It is in the different forms of convergence of random variables that separates the *weak* and the *strong* law of large numbers:

Weak law: $\overline{X}_n \xrightarrow{P} \mu$, for $n \to \infty$.

Strong law: $\overline{X}_n \xrightarrow{\text{a.s.}} \mu$, for $n \to \infty$.

It follows from the LLN that if an event of probability p is observed repeatedly during independent experiments, the ratio of the observed frequency of that event to the total number of repetitions converges towards p.

For example, if $Y_1, Y_2, \ldots$ are independent Bernoulli random variables taking values 1 with probability p and 0 with probability 1-p, then $\mathrm{E}(Y_i) = p$ for all *i*, so that $\overline{Y}_n$ converges to p almost surely.

### Central Limit Theorem

"The central limit theorem (CLT) is one of the great results of mathematics." It explains the ubiquitous occurrence of the normal distribution in nature.

The theorem states that the average of many independent and identically distributed random variables with finite variance tends towards a normal distribution *irrespective* of the distribution followed by the original random variables. Formally, let $X_1, X_2, \ldots$ be independent random variables with mean $\mu$ and variance $\sigma^2 > 0$ Then the sequence of random variables,

$$Z_n = \frac{\sum_{i=1}^{n} (X_i - \mu)}{\sigma\sqrt{n}}$$

converges in distribution to a standard normal random variable.

For some classes of random variables the classic central limit theorem works rather fast, for example the distributions with finite first, second, and third moment from the exponential family; on the other hand, for some random variables of the heavy tail and fat tail variety, it works very slowly or may not work at all: in such cases one may use the Generalized Central Limit Theorem (GCLT).

## Event

In probability theory, an event is a set of outcomes of an experiment (a subset of the sample space) to which a probability is assigned. A single outcome may be an element of many different events, and different events in an experiment are usually not equally likely, since they may include very different groups of outcomes. An event defines a complementary event, namely the complementary set (the event *not* occurring), and together these define a Bernoulli trial: did the event occur or not?

Typically, when the sample space is finite, any subset of the sample space is an event (*i.e.* all elements of the power set of the sample space are defined as events). However, this approach does not work well in cases where the sample space is uncountably infinite. So, when defining a probability space it is possible, and often necessary, to exclude certain subsets of the sample space from being events.

Example:

If we assemble a deck of 52 playing cards with no jokers, and draw a single card from the deck, then the sample space is a 52-element set, as each card is a possible outcome. An event, however, is any subset of the sample space, including any singleton set (an elementary event), the empty set (an impossible event, with probability zero) and the sample space itself (a certain event, with probability one). Other events are proper subsets of the sample space that contain multiple elements. So, for example, potential events include:

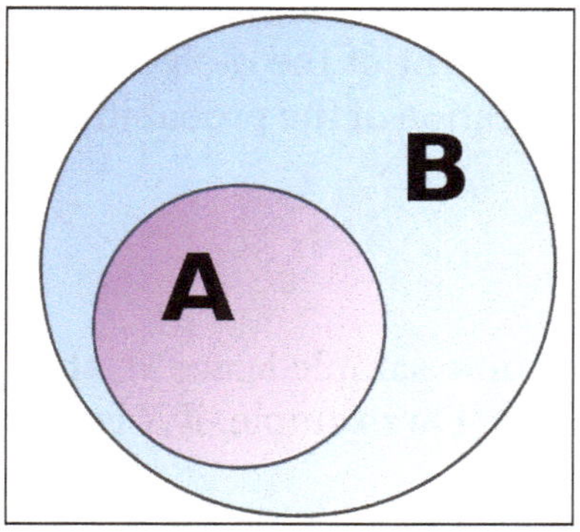


An Euler diagram of an event. *B* is the sample space and *A* is an event. By the ratio of their areas, the probability of *A* is approximately 0.4.

- "Red and black at the same time without being a joker" (0 elements),
- "The 5 of Hearts" (1 element),
- "A King" (4 elements),
- "A Face card" (12 elements),
- "A Spade" (13 elements),
- "A Face card or a red suit" (32 elements),
- "A card" (52 elements).

Since all events are sets, they are usually written as sets (e.g. {1, 2, 3}), and represented graphically using Venn diagrams. In the situation where each outcome in the sample space Ω is equally likely, the probability *P* of an event *A* is the following formula:

$$\mathrm{P}(A) = \frac{|A|}{|\Omega|} \left( \text{alternatively: } \Pr(A) = \frac{|A|}{|\Omega|} \right)$$

This rule can readily be applied to each of the example events above.

## Events in Probability Spaces

Defining all subsets of the sample space as events works well when there are only finitely many outcomes, but gives rise to problems when the sample space is infinite. For many standard probability distributions, such as the normal distribution, the sample space is the set of real numbers or some subset of the real numbers. Attempts to define probabilities for all subsets of the real numbers run into difficulties when one considers 'badly behaved' sets, such as those that are nonmeasurable. Hence, it is necessary to restrict attention to a more limited family of subsets. For the standard tools of probability theory, such as joint and conditional probabilities, to work, it is necessary to use a σ-algebra, that is, a family closed under complementation and countable unions of its members. The most natural choice is the Borel measurable set derived from unions and intersections of intervals. However, the larger class of Lebesgue measurable sets proves more useful in practice.

In the general measure-theoretic description of probability spaces, an event may be defined as an element of a selected σ-algebra of subsets of the sample space. Under this definition, any subset of the sample space that is not an element of the σ-algebra is not an event, and does not have a probability. With a reasonable specification of the probability space, however, all *events of interest* are elements of the σ-algebra.

## Notation

Even though events are subsets of some sample space Ω, they are often written as predicates or indicators involving random variables. For example, if $X$ is a real-valued random variable defined on the sample space Ω, the event,

$$\{\omega \in \Omega \mid u < X(\omega) \leq v\}$$

can be written more conveniently as, simply,

$$u < X \leq v.$$

This is especially common in formulas for a probability, such as:

$$\Pr(u < X \leq v) = F(v) - F(u).$$

The set $u < X \leq v$ is an example of an inverse image under the mapping $X$ because $\omega \in X^{-1}((u,v])$ if and only if $u < X(\omega) \leq v$.

## Elementary Event

In probability theory, an elementary event (also called an atomic event or sample point) is an event which contains only a single outcome in the sample space. Using set theory terminology, an elementary event is a singleton. Elementary events and their corresponding outcomes are often written interchangeably for simplicity, as such an event corresponds to precisely one outcome.

The following are examples of elementary events:

- All sets $\{k\}$, where $k \in \mathrm{N}$ if objects are being counted and the sample space is $S = \{0, 1, 2, 3, \ldots\}$ (the natural numbers).

- {HH}, {HT}, {TH} and {TT} if a coin is tossed twice. $S$ = {HH, HT, TH, TT}. H stands for heads and T for tails.
- All sets $\{x\}$, where $x$ is a real number. Here $X$ is a random variable with a normal distribution and $S = (-\infty, +\infty)$. This example shows that, because the probability of each elementary event is zero, the probabilities assigned to elementary events do not determine a continuous probability distribution.

## Probability of an Elementary Event

Elementary events may occur with probabilities that are between zero and one (inclusively). In a discrete probability distribution whose sample space is finite, each elementary event is assigned a particular probability. In contrast, in a continuous distribution, individual elementary events must all have a probability of zero because there are infinitely many of them— then non-zero probabilities can only be assigned to non-elementary events.

Some "mixed" distributions contain both stretches of continuous elementary events and some discrete elementary events; the discrete elementary events in such distributions can be called atoms or atomic events and can have non-zero probabilities.

Under the measure-theoretic definition of a probability space, the probability of an elementary event need not even be defined. In particular, the set of events on which probability is defined may be some σ-algebra on $S$ and not necessarily the full power set.

## Complementary Event

In probability theory, the complement of any event $A$ is the event [not $A$], i.e. the event that $A$ does not occur. The event $A$ and its complement [not $A$] are mutually exclusive and exhaustive. Generally, there is only one event $B$ such that $A$ and $B$ are both mutually exclusive and exhaustive; that event is the complement of $A$. The complement of an event $A$ is usually denoted as $A'$, $A^c$ or $A$. Given an event, the event and its complementary event define a Bernoulli trial: did the event occur or not?

For example, if a typical coin is tossed and one assumes that it cannot land on its edge, then it can either land showing "heads" or "tails." Because these two outcomes are mutually exclusive (i.e. the coin cannot simultaneously show both heads and tails) and collectively exhaustive (i.e. there are no other possible outcomes not represented between these two), they are therefore each other's complements. This means that [heads] is logically equivalent to [not tails], and [tails] is equivalent to [not heads].

## Complement Rule

In a random experiment, the probabilities of all possible events (the sample space) must total to 1— that is, some outcome must occur on every trial. For two events to be complements, they must be collectively exhaustive, together filling the entire sample space. Therefore, the probability of an event's complement must be unity minus the probability of the event. That is, for an event $A$,

$$P(A^c) = 1 - P(A).$$

Equivalently, the probabilities of an event and its complement must always total to 1. This does not, however, mean that *any* two events whose probabilities total to 1 are each other's complements; complementary events must also fulfill the condition of mutual exclusivity.

### Example of the Utility of this Concept

Suppose one throws an ordinary six-sided die eight times. What is the probability that one sees a "1" at least once?

It may be tempting to say that:

Pr(["1" on 1st trial] or ["1" on second trial] or ... or ["1" on 8th trial])

= Pr("1" on 1st trial) + Pr("1" on second trial) + ... + P("1" on 8th trial)

= 1/6 + 1/6 + ... + 1/6.

= 8/6 = 1.3333... (and this is clearly wrong)

That cannot be right because a probability cannot be more than 1. The technique is wrong because the eight events whose probabilities got added are not mutually exclusive.

One may resolve this overlap by the principle of inclusion-exclusion, or in this case one may instead more simply find the probability of the complementary event and subtract it from 1, thus:

Pr(at least one "1") = 1 − Pr(no "1"s)

= 1 − Pr([no "1" on 1st trial] and [no "1" on 2nd trial] and ... and [no "1" on 8th trial])

= 1 − Pr(no "1" on 1st trial) × Pr(no "1" on 2nd trial) × ... × Pr(no "1" on 8th trial)

$= 1 -(5/6) \times (5/6) \times \ldots \times (5/6)$

$= 1 - (5/6)^8$

= 0.7674.

## Probability Distribution

In probability theory and statistics, a probability distribution is a mathematical function that provides the probabilities of occurrence of different possible outcomes in an experiment. In more technical terms, the probability distribution is a description of a random phenomenon in terms of the probabilities of events. For instance, if the random variable $X$ is used to denote the outcome of a coin toss ("the experiment"), then the probability distribution of $X$ would take the value 0.5 for $X$ = heads, and 0.5 for $X$ = tails (assuming the coin is fair). Examples of random phenomena can include the results of an experiment or survey.

A probability distribution is specified in terms of an underlying sample space, which is the set of all possible outcomes of the random phenomenon being observed. The sample space may be the set of real numbers or a set of vectors, or it may be a list of non-numerical values; for example, the sample space of a coin flip would be {heads, tails} .

Probability distributions are generally divided into two classes. A discrete probability distribution (applicable to the scenarios where the set of possible outcomes is discrete, such as a coin toss or a roll of dice) can be encoded by a discrete list of the probabilities of the outcomes, known as a probability mass function. On the other hand, a continuous probability distribution (applicable to the scenarios where the set of possible outcomes can take on values in a continuous range (e.g. real numbers), such as the temperature on a given day) is typically described by probability density functions (with the probability of any individual outcome actually being 0). The normal distribution is a commonly encountered continuous probability distribution. More complex experiments, such as those involving stochastic processes defined in continuous time, may demand the use of more general probability measures.

A probability distribution whose sample space is one-dimensional (for example real numbers, list of labels, ordered labels or binary) is called univariate, while a distribution whose sample space is a vector space of dimension 2 or more is called multivariate. A univariate distribution gives the probabilities of a single random variable taking on various alternative values; a multivariate distribution (a joint probability distribution) gives the probabilities of a random vector – a list of two or more random variables – taking on various combinations of values. Important and commonly encountered univariate probability distributions include the binomial distribution, the hypergeometric distribution, and the normal distribution. The multivariate normal distribution is a commonly encountered multivariate distribution.

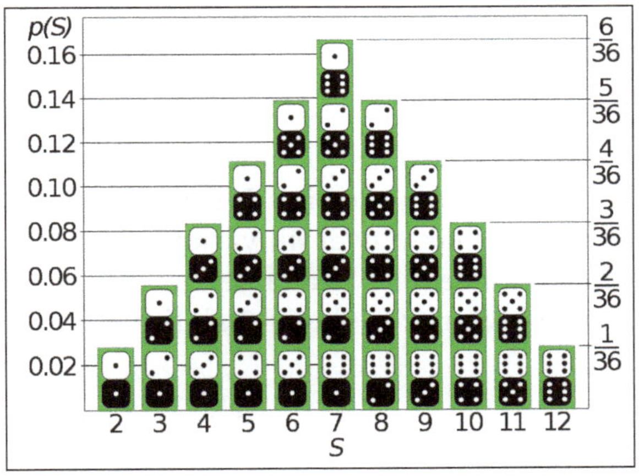


The probability mass function (pmf) $p(S)$ specifies the probability distribution for the sum $S$ of counts from two dice. For example, the figure shows that $p(11) = 2/36 = 1/18$. The pmf allows the computation of probabilities of events such as $P(S > 9) = 1/12 + 1/18 + 1/36 = 1/6$, and all other probabilities in the distribution.

To define probability distributions for the simplest cases, it is necessary to distinguish between discrete and continuous random variables. In the discrete case, it is sufficient to specify a probability mass function $p$ assigning a probability to each possible outcome: for example, when throwing a fair dice, each of the six values 1 to 6 has the probability 1/6. The probability of an event is then defined to be the sum of the probabilities of the outcomes that satisfy the event; for example, the probability of the event "the dice rolls an even value" is:

$$p(2)+p(4)+p(6)=1/6+1/6+1/6=1/2.$$

In contrast, when a random variable takes values from a continuum then typically, any individual outcome has probability zero and only events that include infinitely many outcomes, such as intervals, can have positive probability. For example, the probability that a given object weighs *exactly* 500 g is zero, because the probability of measuring exactly 500 g tends to zero as the accuracy of our measuring instruments increases. Nevertheless, in quality control one might demand that the probability of a "500 g" package containing between 490 g and 510 g should be no less than 98%, and this demand is less sensitive to the accuracy of measurement instruments.

Continuous probability distributions can be described in several ways. The probability density function describes the infinitesimal probability of any given value, and the probability that the outcome lies in a given interval can be computed by integrating the probability density function over that interval. On the other hand, the cumulative distribution function describes the probability that the random variable is no larger than a given value; the probability that the outcome lies in a given interval can be computed by taking the difference between the values of the cumulative distribution function at the endpoints of the interval. The cumulative distribution function is the antiderivative of the probability density function provided that the latter function exists.

As probability theory is used in quite diverse applications, terminology is not uniform and sometimes confusing. The following terms are used for non-cumulative probability distribution functions:

- Frequency distribution: A frequency distribution is a table that displays the frequency of various outcomes in a sample.
- Relative frequency distribution: A frequency distribution where each value has been divided (normalized) by a number of outcomes in a sample i.e. sample size.
- Probability distribution: Sometimes used as an alias for *Relative frequency distribution* but most books use it as a limit to which *Relative frequency distribution* tends when sample size tends to population size. It's a general term to indicate the way the total probability of 1 is distributed over all various possible outcomes (i.e. over entire population). It may for instance refer to a table that displays the probabilities of various outcomes in a finite population or to the probability density of an uncountably infinite population.
- Cumulative distribution function: Is a general functional form to describe a probability distribution.
- Probability distribution function: Somewhat ambiguous term sometimes referring to a functional form of probability distribution table. Could be called a "normalized frequency distribution function", where area under the graph equals to 1.
- Probability mass, Probability mass function, (p.m.f.) Discrete probability distribution function: For discrete random variables.
- Categorical distribution: For discrete random variables with a finite set of values.
- Probability density, Probability density function, (p.d.f.) Continuous probability distribution function: Most often reserved for continuous random variables.

The following terms are somewhat ambiguous as they can refer to non-cumulative or cumulative distributions, depending on authors' preferences:

- Probability distribution function: continuous or discrete, non-cumulative or cumulative.
- Probability function: even more ambiguous, can mean any of the above or other things.

## Cumulative Distribution Function

Because a probability distribution $P$ on the real line is determined by the probability of a scalar random variable $X$ being in a half-open interval $(-\infty, x]$, the probability distribution is completely characterized by its cumulative distribution function:

$$F(x) = \mathrm{P}[X \le x] \qquad \text{for all } x \in \mathbb{R}.$$

## Discrete Probability Distribution

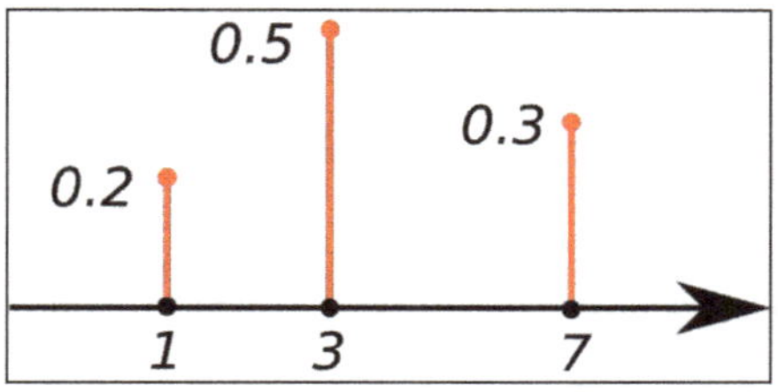


The probability mass function of a discrete probability distribution. The probabilities of the singletons {1}, {3}, and {7} are respectively 0.2, 0.5, 0.3. A set not containing any of these points has probability zero.

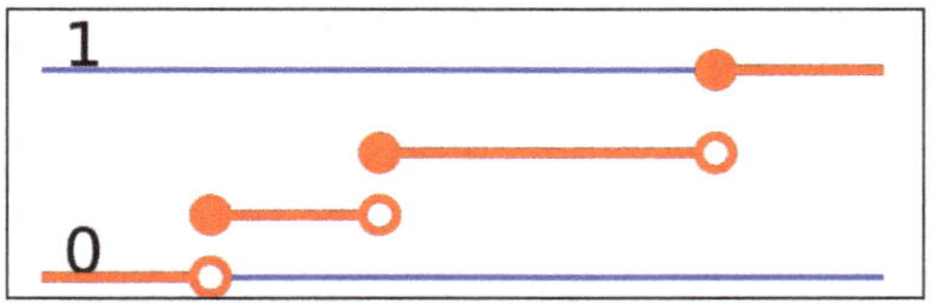


The cdf of a discrete probability distribution.

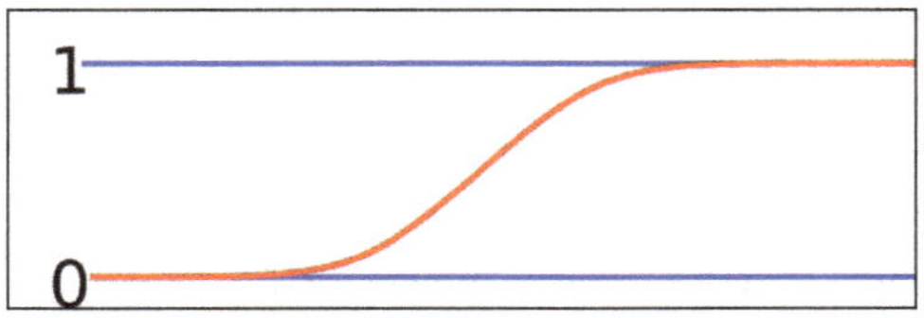


A continuous probability distribution.

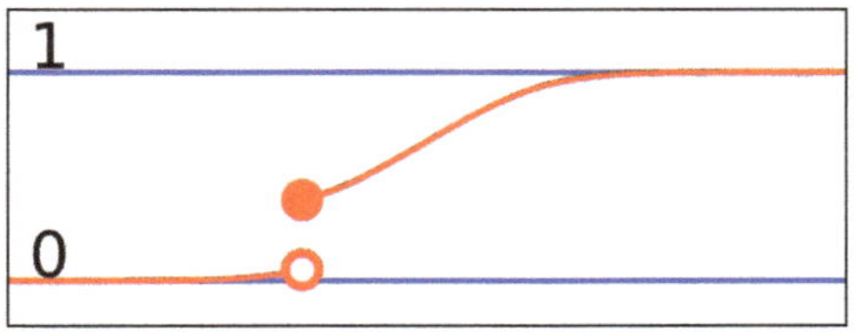


A distribution which has both a continuous part and a discrete part.

A discrete probability distribution is a probability distribution that can take on a countable number of values. For the probabilities to add up to 1, they have to decline to zero fast enough. For example, if $\mathrm{P}(X = n) = \frac{1}{2^n}$ for $n$ = 1, 2, ..., the sum of probabilities would be 1/2 + 1/4 + 1/8 + ... = 1.

Well-known discrete probability distributions used in statistical modeling include the Poisson distribution, the Bernoulli distribution, the binomial distribution, the geometric distribution, and the negative binomial distribution. Additionally, the discrete uniform distribution is commonly used in computer programs that make equal-probability random selections between a number of choices.

When a sample (a set of observations) is drawn from a larger population, the sample points have an empirical distribution that is discrete and that provides information about the population distribution.

## Measure Theoretic Formulation

A measurable function $X: A \to B$ between a probability space $(A, \mathcal{A}, P)$ and a measurable space $(B, \mathcal{B})$ is called a discrete random variable provided that its image is a countable set. In this case measurability of $X$ means that the pre-images of singleton sets are measurable, i.e., $X^{-1}(\{b\}) \in \mathcal{A}$ for all $b \in B$. The latter requirement induces a probability mass function $f_X : X(A) \to \mathbb{R}$ via $f_X(b) := P(X^{-1}(\{b\}))$. Since the pre-images of disjoint sets are disjoint,

$$\sum_{b \in X(A)} f_X(b) = \sum_{b \in X(A)} P(X^{-1}(\{b\})) = P\left( \bigcup_{b \in X(A)} X^{-1}(\{b\}) \right) = P(A) = 1.$$

This recovers the definition given above.

## Cumulative Distribution Function

Equivalently to the above, a discrete random variable can be defined as a random variable whose cumulative distribution function (cdf) increases only by jump discontinuities—that is, its cdf increases only where it "jumps" to a higher value, and is constant between those jumps. Note however that the points where the cdf jumps may form a dense set of the real numbers. The points where jumps occur are precisely the values which the random variable may take.

## Delta-function Representation

Consequently, a discrete probability distribution is often represented as a generalized probability density function involving Dirac delta functions, which substantially unifies the treatment of continuous and discrete distributions. This is especially useful when dealing with probability distributions involving both a continuous and a discrete part.

## Indicator-function Representation

For a discrete random variable $X$, let $u_0$, $u_1$, ... be the values it can take with non-zero probability. Denote:

$$\Omega_i = X^{-1}(u_i) = \{\omega : X(\omega) = u_i\}, i = 0, 1, 2, \ldots$$

These are disjoint sets, and for such sets:

$$P\left( \bigcup_i \Omega_i \right) = \sum_i P(\Omega_i) = \sum_i P(X = u_i) = 1.$$

It follows that the probability that $X$ takes any value except for $u_0$, $u_1$, ... is zero, and thus one can write $X$ as,

$$X(\omega) = \sum_i u_i 1_{\Omega_i}(\omega)$$

except on a set of probability zero, where $1_A$ is the indicator function of $A$. This may serve as an alternative definition of discrete random variables.

## Continuous Probability Distribution

A continuous probability distribution is a probability distribution with a cumulative distribution function that is absolutely continuous. Equivalently, it is a probability distribution on the real numbers that is absolutely continuous with respect to Lebesgue measure. Such distributions can be represented by their probability density functions. If the distribution of $X$ is continuous, then $X$ is called a continuous random variable. There are many examples of continuous probability distributions: normal, uniform, chi-squared, and others.

Formally, if $X$ is a continuous random variable, then it has a probability density function $f(x)$, and therefore its probability of falling into a given interval, say $[a, b]$, is given by the integral:

$$\mathrm{P}[a \le X \le b] = \int_a^b f(x)dx$$

In particular, the probability for $X$ to take any single value $a$ (that is $a \le X \le a$) is zero, because an integral with coinciding upper and lower limits is always equal to zero.

Some authors use the term "continuous distribution" to denote distributions whose cumulative distribution functions are continuous, rather than absolutely continuous. These distributions are the ones $\mu$ such that $\mu\{x\}=0$ for all $x$. This definition includes the (absolutely) continuous distributions defined above, but it also includes singular distributions, which are neither continuous nor discrete nor a mixture of those. An example is given by the Cantor distribution.

## Properties

- The probability distribution of the sum of two independent random variables is the convolution of each of their distributions.
- Probability distributions are not a vector space—they are not closed under linear combinations, as these do not preserve non-negativity or total integral 1—but they are closed under convex combination, thus forming a convex subset of the space of functions (or measures).

## Kolmogorov Definition

In the measure-theoretic formalization of probability theory, a random variable is defined as a measurable function $X$ from a probability space $(\Omega, \mathcal{F}, \mathrm{P})$ to a measurable space $(\mathcal{X}, \mathcal{A})$. A probability distribution of $X$ is the pushforward measure $X_*\mathrm{P}$ of $X$, which is a probability measure on $(\mathcal{X},\mathcal{A})$ satisfying $X_*\mathrm{P} = \mathrm{P}X^{-1}$.

## Random Number Generation

A frequent problem in statistical simulations (the Monte Carlo method) is the generation of pseudo-random numbers that are distributed in a given way. Most algorithms are based on a pseudorandom number generator that produces numbers $X$ that are uniformly distributed in the half-open interval $[0,1)$. These random variates $X$ are then transformed via some algorithm to create a new random variate having the required probability distribution.

## Applications

The concept of the probability distribution and the random variables which they describe underlies the mathematical discipline of probability theory, and the science of statistics. There is spread or variability in almost any value that can be measured in a population (e.g. height of people, durability of a metal, sales growth, traffic flow, etc.); almost all measurements are made with some intrinsic error; in physics many processes are described probabilistically, from the kinetic properties of gases to the quantum mechanical description of fundamental particles. For these and many other reasons, simple numbers are often inadequate for describing a quantity, while probability distributions are often more appropriate.

Several more specific example of an application:

- The cache language models and other statistical language models used in natural language processing to assign probabilities to the occurrence of particular words and word sequences do so by means of probability distributions.
- Probabilistic load flow in power-flow study explains the uncertainties of input variables as probability distribution and provide the power flow calculation also in term of probability distribution.

## Common Probability Distributions

The following is a list of some of the most common probability distributions, grouped by the type of process that they are related to.

All of the univariate distributions below are singly peaked; that is, it is assumed that the values cluster around a single point. In practice, actually observed quantities may cluster around multiple values. Such quantities can be modeled using a mixture distribution.

### Related to Real-valued Quantities that Grow Linearly (e.g. Errors, Offsets)

- Normal distribution (Gaussian distribution), for a single such quantity; the most commonly used continuous distribution.

### Related to Positive Real-valued Quantities that Grow Exponentially (e.g. Prices, Incomes, Populations)

- Log-normal distribution, for a single such quantity whose log is normally distributed.
- Pareto distribution, for a single such quantity whose log is exponentially distributed; the prototypical power law distribution.

### Related to Real-valued Quantities that are Assumed to be Uniformly Distributed over a (Possibly Unknown) Region

- Discrete uniform distribution, for a finite set of values (e.g. the outcome of a fair die).
- Continuous uniform distribution, for continuously distributed values.

### Related to Bernoulli Trials (Yes/No Events, with a Given Probability)

- Basic distributions:
    - Bernoulli distribution, for the outcome of a single Bernoulli trial (e.g. success/failure, yes/no).
    - Binomial distribution, for the number of "positive occurrences" (e.g. successes, yes votes, etc.) given a fixed total number of independent occurrences.
    - Negative binomial distribution, for binomial-type observations but where the quantity of interest is the number of failures before a given number of successes occurs.
    - Geometric distribution, for binomial-type observations but where the quantity of interest is the number of failures before the first success; a special case of the negative binomial distribution.
- Related to sampling schemes over a finite population:
    - Hypergeometric distribution, for the number of "positive occurrences" (e.g. successes, yes votes, etc.) given a fixed number of total occurrences, using sampling without replacement.
    - Beta-binomial distribution, for the number of "positive occurrences" (e.g. successes, yes votes, etc.) given a fixed number of total occurrences, sampling using a Pólya urn scheme (in some sense, the "opposite" of sampling without replacement).

### Related to Categorical Outcomes (Events with *K* Possible Outcomes, with a Given Probability for each Outcome)

- Categorical distribution, for a single categorical outcome (e.g. yes/no/maybe in a survey); a generalization of the Bernoulli distribution.
- Multinomial distribution, for the number of each type of categorical outcome, given a fixed number of total outcomes; a generalization of the binomial distribution.

- Multivariate hypergeometric distribution, similar to the multinomial distribution, but using sampling without replacement; a generalization of the hypergeometric distribution.

### Related to Events in a Poisson Process (Eventsthat Occur Independently with a Given Rate)

- Poisson distribution, for the number of occurrences of a Poisson-type event in a given period of time.
- Exponential distribution, for the time before the next Poisson-type event occurs.
- Gamma distribution, for the time before the next k Poisson-type events occur.

### Related to the Absolute Values of Vectors with Normally Distributed Components

- Rayleigh distribution, for the distribution of vector magnitudes with Gaussian distributed orthogonal components. Rayleigh distributions are found in RF signals with Gaussian real and imaginary components.
- Rice distribution, a generalization of the Rayleigh distributions for where there is a stationary background signal component. Found in Rician fading of radio signals due to multipath propagation and in MR images with noise corruption on non-zero NMR signals.

### Related to Normally Distributed Quantities Operated with Sum of Squares (For Hypothesis Testing)

- Chi-squared distribution, the distribution of a sum of squared standard normal variables; useful e.g. for inference regarding the sample variance of normally distributed samples.
- Student's t distribution, the distribution of the ratio of a standard normal variable and the square root of a scaled chi squared variable; useful for inference regarding the mean of normally distributed samples with unknown variance.
- F-distribution, the distribution of the ratio of two scaled chi squared variables; useful e.g. for inferences that involve comparing variances or involving R-squared (the squared correlation coefficient).

### Useful as Conjugate Prior Distributions in Bayesian Inference

- Beta distribution, for a single probability (real number between 0 and 1); conjugate to the Bernoulli distribution and binomial distribution.
- Gamma distribution, for a non-negative scaling parameter; conjugate to the rate parameter of a Poisson distribution or exponential distribution, the precision (inverse variance) of a normal distribution, etc.
- Dirichlet distribution, for a vector of probabilities that must sum to 1; conjugate to the categorical distribution and multinomial distribution; generalization of the beta distribution.

- Wishart distribution, for a symmetric non-negative definite matrix; conjugate to the inverse of the covariance matrix of a multivariate normal distribution; generalization of the gamma distribution.

# Binomial Distribution

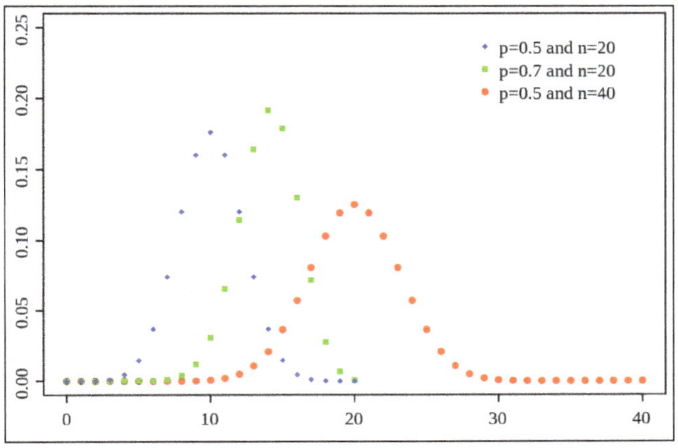


Cumulative distribution function.

In probability theory and statistics, the binomial distribution with parameters $n$ and $p$ is the discrete probability distribution of the number of successes in a sequence of $n$ independent experiments, each asking a yes–no question, and each with its own boolean-valued outcome: success/yes/true/one (with probability $p$) or failure/no/false/zero (with probability $q = 1 - p$). A single success/failure experiment is also called a Bernoulli trial or Bernoulli experiment and a sequence of outcomes is called a Bernoulli process; for a single trial, i.e., $n = 1$, the binomial distribution is a Bernoulli distribution. The binomial distribution is the basis for the popular binomial test of statistical significance.

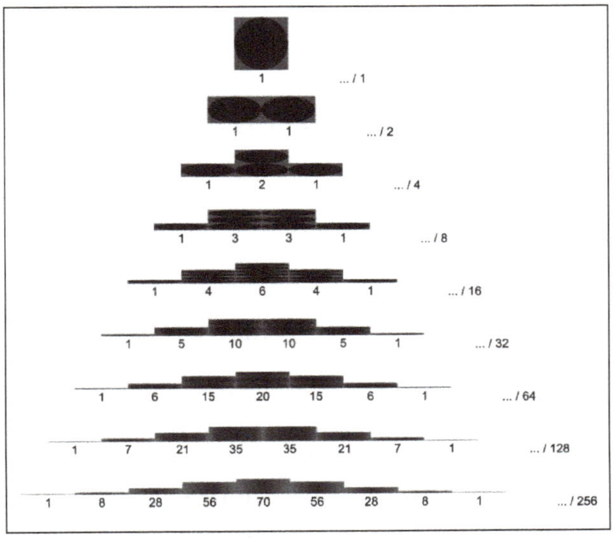


Binomial distribution for $p = 0.5$ with $n$ and $k$ as in Pascal's triangle The probability that a ball in a Galton box with 8 layers ($n = 8$) ends up in the central bin ($k = 4$) is $70/256$.

The binomial distribution is frequently used to model the number of successes in a sample of size $n$ drawn with replacement from a population of size $N$. If the sampling is carried out without replacement, the draws are not independent and so the resulting distribution is a hypergeometric distribution, not a binomial one. However, for $N$ much larger than $n$, the binomial distribution remains a good approximation, and is widely used.

## Specification

### Probability Mass Function

In general, if the random variable $X$ follows the binomial distribution with parameters $n \in \mathrm{N}$ and $p \in [0,1]$, we write $X \sim \mathrm{B}(n, p)$. The probability of getting exactly $k$ successes in $n$ trials is given by the probability mass function:

$$f(k,n,p) = \Pr(k;n,p) = \Pr(X=k) = \binom{n}{k} p^k (1-p)^{n-k}$$

for $k = 0, 1, 2, \ldots, n$, where,

$$\binom{n}{k} = \frac{n!}{k!(n-k)!}$$

is the binomial coefficient, hence the name of the distribution. The formula can be understood as follows. $k$ successes occur with probability $p^k$ and $n - k$ failures occur with probability $(1 - p)^{n-k}$. However, the $k$ successes can occur anywhere among the $n$ trials, and there are $\binom{n}{k}$ different ways of distributing $k$ successes in a sequence of $n$ trials.

In creating reference tables for binomial distribution probability, usually the table is filled in up to $n/2$ values. This is because for $k > n/2$, the probability can be calculated by its complement as:

$$f(k,n,p) = f(n-k,n,1-p).$$

Looking at the expression $f(k, n, p)$ as a function of $k$, there is a $k$ value that maximizes it. This $k$ value can be found by calculating,

$$\frac{f(k+1,n,p)}{f(k,n,p)} = \frac{(n-k)p}{(k+1)(1-p)}$$

and comparing it to 1. There is always an integer $M$ that satisfies,

$$(n+1)p - 1 \le M < (n+1)p.$$

$f(k, n, p)$ is monotone increasing for $k < M$ and monotone decreasing for $k > M$, with the exception of the case where $(n + 1)p$ is an integer. In this case, there are two values for which $f$ is maximal: $(n + 1)p$ and $(n + 1)p - 1$. $M$ is the *most probable* outcome (that is, the most likely, although this can still be unlikely overall) of the Bernoulli trials and is called the mode.

## Cumulative Distribution Function

The cumulative distribution function can be expressed as:

$$F(k;n,p) = \Pr(X \le k) = \sum_{i=0}^{\lfloor k \rfloor} \binom{n}{i} p^i (1-p)^{n-i}$$

where $\lfloor k \rfloor$ is the "floor" under $k$, i.e. the greatest integer less than or equal to $k$.

It can also be represented in terms of the regularized incomplete beta function, as follows:

$$\begin{aligned} F(k;n,p) &= \Pr(X \le k) \\ &= I_{1-p}(n-k, k+1) \\ &= (n-k)\binom{n}{k}\int_0^{1-p} t^{n-k-1}(1-t)^k \, dt. \end{aligned}$$

Some closed-form bounds for the cumulative distribution function are given below.

Example: Suppose a biased coin comes up heads with probability 0.3 when tossed. What is the probability of achieving 0, 1,..., 6 heads after six tosses?

$$\Pr(0 \text{ heads}) = f(0) = \Pr(X=0) = \binom{6}{0} 0.3^0 (1-0.3)^{6-0} = 0.117649$$

$$\Pr(1 \text{ heads}) = f(1) = \Pr(X=1) = \binom{6}{1} 0.3^1 (1-0.3)^{6-1} = 0.302526$$

$$\Pr(2 \text{ heads}) = f(2) = \Pr(X=2) = \binom{6}{2} 0.3^2 (1-0.3)^{6-2} = 0.324135$$

$$\Pr(3 \text{ heads}) = f(3) = \Pr(X=3) = \binom{6}{3} 0.3^3 (1-0.3)^{6-3} = 0.18522$$

$$\Pr(4 \text{ heads}) = f(4) = \Pr(X=4) = \binom{6}{4} 0.3^4 (1-0.3)^{6-4} = 0.059535$$

$$\Pr(5 \text{ heads}) = f(5) = \Pr(X=5) = \binom{6}{5} 0.3^5 (1-0.3)^{6-5} = 0.010206$$

$$\Pr(6 \text{ heads}) = f(6) = \Pr(X=6) = \binom{6}{6} 0.3^6 (1-0.3)^{6-6} = 0.000729$$

## Expectation

If $X \sim B(n, p)$, that is, $X$ is a binomially distributed random variable, n being the total number of experiments and p the probability of each experiment yielding a successful result, then the expected value of $X$ is:

$$\mathrm{E}[X] = np.$$

For example, if $n = 100$, and $p = 1/4$, then the average number of successful results will be 25.

We calculate the mean, $\mu$, directly calculated from its definition,

$$\mu = \sum_{i=0}^{n} x_i p_i,$$

and the binomial theorem:

$$\begin{aligned}
\mu &= \sum_{k=0}^{n} k \binom{n}{k} p^k (1-p)^{n-k} \\
&= np \sum_{k=0}^{n} k \frac{(n-1)!}{(n-k)!k!} p^{k-1} (1-p)^{(n-1)-(k-1)} \\
&= np \sum_{k=1}^{n} \frac{(n-1)!}{((n-1)-(k-1))!(k-1)!} p^{k-1} (1-p)^{(n-1)-(k-1)} \\
&= np \sum_{k=1}^{n} \binom{n-1}{k-1} p^{k-1} (1-p)^{(n-1)-(k-1)} \\
&= np \sum_{\ell=0}^{n-1} \binom{n-1}{\ell} p^{\ell} (1-p)^{(n-1)-\ell} \text{ with } \ell := k-1 \\
&= np \sum_{\ell=0}^{m} \binom{m}{\ell} p^{\ell} (1-p)^{m-\ell} \text{ with } m := n-1 \\
&= np(p+(1-p))^m \\
&= np
\end{aligned}$$

It is also possible to deduce the mean from the equation $X = X_1 + \cdots + X_n$ whereby all $X_i$ are Bernoulli distributed random variables with $E[X_i] = p$ ($X_i = 1$ if the $i$th experiment succeeds and $X_i = 0$ otherwise). We get: $E[X] = E[X_1 + \cdots + X_n] = E[X_1] + \cdots + E[X_n] = \underbrace{p + \cdots + p}_{n \text{ times}} = np$.

## Variance

The variance is:

$$\operatorname{Var}(X) = np(1-p).$$

Proof: Let $X = X_1 + \cdots + X_n$ where all $X_i$ are independently Bernoulli distributed random variables. Since $\operatorname{Var}(X_i) = p(1-p)$, we get:

$$\operatorname{Var}(X) = \operatorname{Var}(X_1 + \cdots + X_n) = \operatorname{Var}(X_1) + \cdots + \operatorname{Var}(X_n) = n\operatorname{Var}(X_1) = np(1-p).$$

## Higher Moments

The first 6 central moments are given by:

$$\begin{aligned}
\mu_1 &= 0, \\
\mu_2 &= np(1-p), \\
\mu_3 &= np(1-p)(1-2p),
\end{aligned}$$

$$\mu_4 = np(1-p)(1+(3n-6)p(1-p)),$$
$$\mu_5 = np(1-p)(1-2p)(1+(10n-12)p(1-p)),$$
$$\mu_6 = np(1-p)(1-30p(1-p)(1-4p(1-p))+5np(1-p)(5-26p(1-p))+15n^2p^2(1-p)^2).$$

## Mode

Usually the mode of a binomial *B*(*n*,*p*) distribution is equal to , where is the floor function. However, when (*n* + 1)*p* is an integer and *p* is neither 0 nor 1, then the distribution has two modes: (*n* + 1)*p* and (*n* + 1)*p* − 1. When *p* is equal to 0 or 1, the mode will be 0 and *n* correspondingly. These cases can be summarized as follows:

$$\text{mode} = \begin{cases} \lfloor (n+1)p \rfloor & \text{if } (n+1)p \text{ is 0 or a noninteger,} \\ (n+1)p \ \text{ and } \ (n+1)p-1 & \text{if } (n+1)p \in \{1,\ldots,n\}, \\ n & \text{if } (n+1)p = n+1. \end{cases}$$

Let,

$$f(k) = \binom{n}{k} p^k q^{n-k}.$$

For $p=0$ only $f(0)$ has a nonzero value with $f(0)=1$. For $p=1$ we find $f(n)=1$ and $f(k)=0$ for $k \neq n$. This proves that the mode is 0 for $p=0$ and $n$ $p=1$ for $p=1$.

Let, $0<p<1$. We find:

$$\frac{f(k+1)}{f(k)} = \frac{(n-k)p}{(k+1)(1-p)}.$$

From this follows:

$$k > (n+1)p-1 \Rightarrow f(k+1) < f(k)$$
$$k = (n+1)p-1 \Rightarrow f(k+1) = f(k)$$
$$k < (n+1)p-1 \Rightarrow f(k+1) > f(k)$$

So when $(n+1)p-1$ is an integer, then $(n+1)p-1$ and $(n+1)p$ is a mode. In the case that $(n+1)p-1 \notin \mathbb{Z}$, then only $\lfloor (n+1)p-1 \rfloor + 1 = \lfloor (n+1)p \rfloor$ is a mode.

## Median

In general, there is no single formula to find the median for a binomial distribution, and it may even be non-unique. However several special results have been established:

- If *np* is an integer, then the mean, median, and mode coincide and equal *np*.
- Any median *m* must lie within the interval $\lfloor np \rfloor \le m \le \lceil np \rceil$.

- A median $m$ cannot lie too far away from the mean: $|m - np| \le \min\{\ln 2, \max\{p, 1 - p\}\}$.
- The median is unique and equal to $m = \text{round}(np)$ when $|m - np| \le \min\{p, 1 - p\}$ (except for the case when $p = 1/2$ and $n$ is odd).
- When $p = 1/2$ and $n$ is odd, any number $m$ in the interval $1/2(n - 1) \le m \le 1/2(n + 1)$ is a median of the binomial distribution. If $p = 1/2$ and $n$ is even, then $m = n/2$ is the unique median.

## Covariance between two Binomials

If two binomially distributed random variables $X$ and $Y$ are observed together, estimating their covariance can be useful. The covariance is:

$$\text{Cov}(X,Y) = \text{E}(XY) - \mu_X \mu_Y.$$

In the case $n = 1$ (the case of Bernoulli trials) $XY$ is non-zero only when both $X$ and $Y$ are one, and $\mu_X$ and $\mu_Y$ are equal to the two probabilities. Defining $p_B$ as the probability of both happening at the same time, this gives,

$$\text{Cov}(X,Y) = p_B - p_X p_Y,$$

and for $n$ independent pairwise trials:

$$\text{Cov}(X,Y)_n = n(p_B - p_X p_Y).$$

If $X$ and $Y$ are the same variable, this reduces to the variance formula given above.

## Related Distributions

### Sums of Binomials

If $X \sim \text{B}(n, p)$ and $Y \sim \text{B}(m, p)$ are independent binomial variables with the same probability $p$, then $X + Y$ is again a binomial variable; its distribution is $Z=X+Y \sim \text{B}(n+m, p)$:

$$\begin{aligned} \text{P}(Z = k) &= \sum_{i=0}^{k}\left[\binom{n}{i} p^i (1-p)^{n-i}\right]\left[\binom{m}{k-i} p^{k-i} (1-p)^{m-k+i}\right] \\ &= \binom{n+m}{k} p^k (1-p)^{n+m-k} \end{aligned}$$

However, if $X$ and $Y$ do not have the same probability $p$, then the variance of the sum will be smaller than the variance of a binomial variable distributed as $B(n+m, \bar{p})$.

### Ratio of two Binomial Distributions

This result was first derived by Katz et al. in 1978.

Let $X \sim \text{B}(n,p_1)$ and $Y \sim \text{B}(m,p_2)$ be independent. Let $T = (X/n)/(Y/m)$.

Then $\log(T)$ is approximately normally distributed with mean $\log(p_1/p_2)$ and variance $((1/p_1) - 1)/n + ((1/p_2) - 1)/m$.

## Conditional Binomials

If $X \sim B(n, p)$ and $Y \mid X \sim B(X, q)$ (the conditional distribution of $Y$, given $X$), then $Y$ is a simple binomial random variable with distribution $Y \sim B(n, pq)$.

For example, imagine throwing $n$ balls to a basket $U_X$ and taking the balls that hit and throwing them to another basket $U_Y$. If $p$ is the probability to hit $U_X$ then $X \sim B(n, p)$ is the number of balls that hit $U_X$. If $q$ is the probability to hit $U_Y$ then the number of balls that hit $U_Y$ is $Y \sim B(X, q)$ and therefore $Y \sim B(n, pq)$.

Proof: Since $X \sim B(n,p)$ and $Y \sim B(X,q)$, by the law of total probability,

$$\begin{aligned}\Pr[Y=m] &= \sum_{k=m}^{n} \Pr[Y=m \mid X=k]\Pr[X=k] \\ &= \sum_{k=m}^{n} \binom{n}{k}\binom{k}{m} p^k q^m (1-p)^{n-k}(1-q)^{k-m}\end{aligned}$$

Since $\binom{n}{k}\binom{k}{m} = \binom{n}{m}\binom{n-m}{k-m}$, the equation above can be expressed as:

$$\Pr[Y=m] = \sum_{k=m}^{n} \binom{n}{m}\binom{n-m}{k-m} p^k q^m (1-p)^{n-k}(1-q)^{k-m}$$

Factoring $p^k = p^m p^{k-m}$ and pulling all the terms that don't depend on $k$ out of the sum now yields:

$$\begin{aligned}\Pr[Y=m] &= \binom{n}{m} p^m q^m \left(\sum_{k=m}^{n} \binom{n-m}{k-m} p^{k-m}(1-p)^{n-k}(1-q)^{k-m}\right) \\ &= \binom{n}{m}(pq)^m \left(\sum_{k=m}^{n} \binom{n-m}{k-m} \left(p(1-q)\right)^{k-m}(1-p)^{n-k}\right)\end{aligned}$$

After substituting $i = k-m$ in the expression above, we get:

$$\Pr[Y=m] = \binom{n}{m}(pq)^m \left(\sum_{i=0}^{n-m} \binom{n-m}{i} (p-pq)^i (1-p)^{n-m-i}\right)$$

Notice that the sum (in the parentheses) above equals $(p-pq+1-p)^{n-m}$ by the binomial theorem. Substituting this in finally yields,

$$\begin{aligned}\Pr[Y=m] &= \binom{n}{m}(pq)^m (p-pq+1-p)^{n-m} \\ &= \binom{n}{m}(pq)^m (1-pq)^{n-m}\end{aligned}$$

and thus $Y \sim B(n, pq)$ as desired.

## Bernoulli Distribution

The Bernoulli distribution is a special case of the binomial distribution, where $n = 1$. Symbolically, $X \sim B(1, p)$ has the same meaning as $X \sim \text{Bernoulli}(p)$. Conversely, any binomial distribution,

B($n$, $p$), is the distribution of the sum of $n$ Bernoulli trials, Bernoulli($p$), each with the same probability $p$.

## Poisson Binomial Distribution

The binomial distribution is a special case of the Poisson binomial distribution, or general binomial distribution, which is the distribution of a sum of $n$ independent non-identical Bernoulli trials B($p_i$).

## Normal Approximation

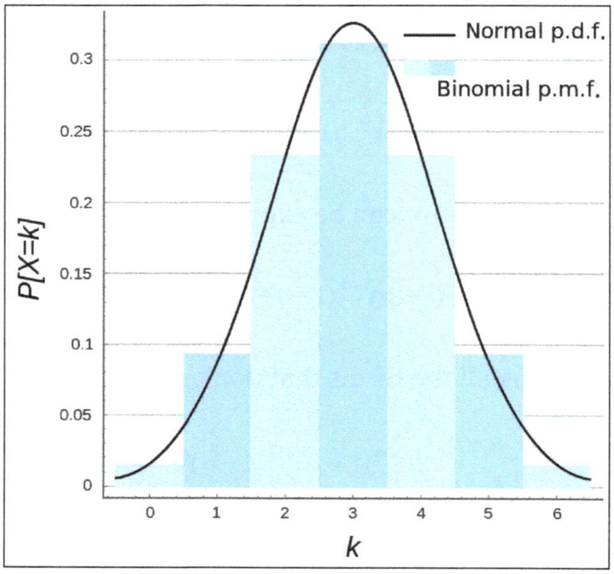


Binomial probability mass function and normal probability density function approximation for $n = 6$ and $p = 0.5$.

If $n$ is large enough, then the skew of the distribution is not too great. In this case a reasonable approximation to B($n$, $p$) is given by the normal distribution,

$$\mathcal{N}(np, np(1-p)),$$

and this basic approximation can be improved in a simple way by using a suitable continuity correction. The basic approximation generally improves as $n$ increases (at least 20) and is better when $p$ is not near to 0 or 1. Various rules of thumb may be used to decide whether $n$ is large enough, and $p$ is far enough from the extremes of zero or one:

- One rule is that for $n > 5$ the normal approximation is adequate if the absolute value of the skewness is strictly less than 1/3; that is, if:

$$\frac{|1-2p|}{\sqrt{np(1-p)}} = \frac{1}{\sqrt{n}}\left|\sqrt{\frac{1-p}{p}} - \sqrt{\frac{p}{1-p}}\right| < \frac{1}{3}.$$

- A stronger rule states that the normal approximation is appropriate only if everything within 3 standard deviations of its mean is within the range of possible values; that is, only if,

$$\mu \pm 3\sigma = np \pm 3\sqrt{np(1-p)} \in (0,n).$$

This 3-standard-deviation rule is equivalent to the following conditions, which also imply the first rule.

$$n > 9\left(\frac{1-p}{p}\right) \quad \text{and} \quad n > 9\left(\frac{p}{1-p}\right).$$

The rule $np \pm 3\sqrt{np(1-p)} \in (0,n)$ is totally equivalent to request that:

$$np - 3\sqrt{np(1-p)} > 0 \quad \text{and} \quad np + 3\sqrt{np(1-p)} < n.$$

Moving terms around yields:

$$np > 3\sqrt{np(1-p)} \quad \text{and} \quad n(1-p) > 3\sqrt{np(1-p)}.$$

Since $0 < p < 1,$ we can apply the square power and divide by the respective factors $np^2$ and $n(1-p)^2$, to obtain the desired conditions:

$$n > 9\left(\frac{1-p}{p}\right) \quad \text{and} \quad n > 9\left(\frac{p}{1-p}\right).$$

Notice that these conditions automatically imply that $n > 9$. On the other hand, apply again the square root and divide by 3,

$$\frac{\sqrt{n}}{3} > \sqrt{\frac{1-p}{p}} > 0 \quad \text{and} \quad \frac{\sqrt{n}}{3} > \sqrt{\frac{p}{1-p}} > 0.$$

Subtracting the second set of inequalities from the first one yields,

$$\frac{\sqrt{n}}{3} > \sqrt{\frac{1-p}{p}} - \sqrt{\frac{p}{1-p}} > -\frac{\sqrt{n}}{3};$$

and so, the desired first rule is satisfied,

$$\left|\sqrt{\frac{1-p}{p}} - \sqrt{\frac{p}{1-p}}\right| < \frac{\sqrt{n}}{3}.$$

Another commonly used rule is that both values $np$ and $n(1-p)$ must be greater than or equal to 5. However, the specific number varies from source to source, and depends on how good an approximation one wants. In particular, if one uses 9 instead of 5, the rule implies the results stated in the previous paragraphs.

Assume that both values $np$ and $n(1-p)$ are greater than 9. Since $0 < p < 1$, we easily have that:

$$np \geq 9 > 9(1-p) \quad \text{and} \quad n(1-p) \geq 9 > 9p.$$

We only have to divide now by the respective factors $p$ and $1-p$, to deduce the alternative form of the 3-standard-deviation rule:

$$n > 9\left(\frac{1-p}{p}\right) \quad \text{and} \quad n > 9\left(\frac{p}{1-p}\right).$$

The following is an example of applying a continuity correction. Suppose one wishes to calculate $\Pr(X \leq 8)$ for a binomial random variable $X$. If $Y$ has a distribution given by the normal approximation, then $\Pr(X \leq 8)$ is approximated by $\Pr(Y \leq 8.5)$. The addition of 0.5 is the continuity correction; the uncorrected normal approximation gives considerably less accurate results.

This approximation, known as de Moivre–Laplace theorem, is a huge time-saver when undertaking calculations by hand (exact calculations with large $n$ are very onerous). Nowadays, it can be seen as a consequence of the central limit theorem since $B(n, p)$ is a sum of $n$ independent, identically distributed Bernoulli variables with parameter $p$. This fact is the basis of a hypothesis test, a "proportion z-test", for the value of $p$ using $x/n$, the sample proportion and estimator of $p$, in a common test statistic.

For example, suppose one randomly samples $n$ people out of a large population and ask them whether they agree with a certain statement. The proportion of people who agree will of course depend on the sample. If groups of $n$ people were sampled repeatedly and truly randomly, the proportions would follow an approximate normal distribution with mean equal to the true proportion $p$ of agreement in the population and with standard deviation $\sigma = \sqrt{\frac{p(1-p)}{n}}$.

## Poisson Approximation

The binomial distribution converges towards the Poisson distribution as the number of trials goes to infinity while the product $np$ remains fixed or at least $p$ tends to zero. Therefore, the Poisson distribution with parameter $\lambda = np$ can be used as an approximation to $B(n, p)$ of the binomial distribution if $n$ is sufficiently large and $p$ is sufficiently small. According to two rules of thumb, this approximation is good if $n \geq 20$ and $p \leq 0.05$, or if $n \geq 100$ and $np \leq 10$.

## Limiting Distributions

- *Poisson limit theorem*: As $n$ approaches $\infty$ and $p$ approaches 0 with the product $np$ held fixed, the Binomial($n$, $p$) distribution approaches the Poisson distribution with expected value $\lambda = np$.
- *de Moivre–Laplace theorem*: As $n$ approaches $\infty$ while $p$ remains fixed, the distribution of approaches the normal distribution with expected value 0 and variance 1. This result is sometimes loosely stated by saying that the distribution of $X$ is asymptotically normal with expected value $np$ and variance $np(1 - p)$. This result is a specific case of the central limit theorem.

$$\frac{X - np}{\sqrt{np(1-p)}}$$

## Beta Distribution

The binomial distribution and beta distribution are different views of the same model of repeated Bernoulli trials. The binomial distribution is the PMF of $k$ successes given $n$ independent events each with a probability $p$ of success. The beta distribution is the PDF for the probability of success

$p$ given $n$ independent events with $k$ observed successes. Mathematically, when $\alpha=k+1$ and $\beta=n-k+1$, the beta distribution and the binomial distribution are related by a factor of $n+1$:

$$Beta(p;\alpha;\beta) = (n+1)Binom(k;n;p)$$

Beta distributions also provide a family of prior probability distributions for binomial distributions in Bayesian inference:

$$P(p;\alpha,\beta) = \frac{p^{\alpha-1}(1-p)^{\beta-1}}{\mathrm{B}(\alpha,\beta)}.$$

## Confidence Intervals

Even for quite large values of $n$, the actual distribution of the mean is significantly nonnormal. Because of this problem several methods to estimate confidence intervals have been proposed.

In the equations for confidence intervals below, the variables have the following meaning:

- $n_1$ is the number of successes out of $n$, the total number of trials.
- $\hat{p} = \frac{n_1}{n}$ is the proportion of successes.
- $z$ is the $1-\frac{1}{2}\alpha$ quantile of a standard normal distribution (i.e., probit) corresponding to the target error rate $\alpha$. For example, for a 95% confidence level the error $\alpha$ = 0.05, so $1-\frac{1}{2}\alpha$ = 0.975 and $z$ = 1.96.

### Wald Method

$$\hat{p} \pm z\sqrt{\frac{\hat{p}(1-\hat{p})}{n}}$$

A continuity correction of $0.5/n$ may be added.

### Agresti–Coull Method

$$\tilde{p} \pm z\sqrt{\frac{\tilde{p}(1-\tilde{p})}{n+z^2}}.$$

Here the estimate of $p$ is modified to:

$$\tilde{p} = \frac{n_1 + \frac{1}{2}z^2}{n+z^2}$$

### Arcsine Method

$$\sin^2\left(\arcsin\left(\sqrt{\hat{p}}\right) \pm \frac{z}{2\sqrt{n}}\right)$$

### Wilson Method

The notation in the formula below differs from the previous formulas in two respects:

- Firstly, $z_x$ has a slightly different interpretation in the formula below: it has its ordinary meaning of 'the $x$th quantile of the standard normal distribution', rather than being a shorthand for 'the $(1 - x)$-th quantile'.
- Secondly, this formula does not use a plus-minus to define the two bounds. Instead, one may use $z = z_{\alpha/2}$ to get the lower bound, or use $z = z_{1-\alpha/2}$ to get the upper bound. For example: for a 95% confidence level the error $\alpha$ = 0.05, so one gets the lower bound by using $z = z_{\alpha/2} = z_{0.025} = -1.96$, and one gets the upper bound by using $z = z_{1-\alpha/2} = z_{0.975} = 1.96$.

$$\frac{\hat{p} + \frac{z^2}{2n} + z\sqrt{\frac{\hat{p}(1-\hat{p})}{n} + \frac{z^2}{4n^2}}}{1 + \frac{z^2}{n}}$$

### Generating Binomial Random Variates

Methods for random number generation where the marginal distribution is a binomial distribution are well-established.

One way to generate random samples from a binomial distribution is to use an inversion algorithm. To do so, one must calculate the probability that $\Pr(X = k)$ for all values $k$ from 0 through $n$. (These probabilities should sum to a value close to one, in order to encompass the entire sample space.) Then by using a pseudorandom number generator to generate samples uniformly between 0 and 1, one can transform the calculated samples into discrete numbers by using the probabilities calculated in the first step.

### Tail Bounds

For $k \leq np$, upper bounds for the lower tail of the distribution function can be derived. Recall that $F(k;n,p) = \Pr(X \leq k)$, the probability that there are at most $k$ successes.

Hoeffding's inequality yields the bound,

$$F(k;n,p) \leq \exp\left(-2\frac{(np-k)^2}{n}\right),$$

and Chernoff's inequality can be used to derive the bound:

$$F(k;n,p) \leq \exp\left(-\frac{1}{2p}\frac{(np-k)^2}{n}\right).$$

Moreover, these bounds are reasonably tight when $p = 1/2$, since the following expression holds for all $k \geq 3n/8$,

$$F(k;n,\tfrac{1}{2}) \le \frac{14}{15}\exp\left(-\frac{16(\frac{n}{2}-k)^2}{n}\right).$$

However, the bounds do not work well for extreme values of $p$. In particular, as $p$ 1, value $F(k;n,p)$ goes to zero (for fixed $k$, $n$ with $k < n$) while the upper bound above goes to a positive constant. In this case a better bound is given by,

$$F(k;n,p) \le \exp\left(-nD\left(\frac{k}{n} \,||\, p\right)\right), \text{ if } 0 < \frac{k}{n} < p$$

where $D(a \,||\, p)$ is the relative entropy between an $a$-coin and a $p$-coin (i.e. between the Bernoulli($a$) and Bernoulli($p$) distribution),

$$D(a \,||\, p) = (a)\log\frac{a}{p} + (1-a)\log\frac{1-a}{1-p}.$$

An equivalent formulation of the bound is:

$$\Pr(X \ge k) = F(n-k;n,1-p) \le \exp\left(-nD\left(\frac{k}{n} \,||\, p\right)\right), \text{ if } 0 < \frac{k}{n} < p$$

Both these bounds are derived directly from the Chernoff bound. It can also be shown that,

$$\Pr(X \ge k) = F(n-k;n,1-p) \ge \frac{1}{(n+1)^2}\exp\left(-nD\left(\frac{k}{n} \,||\, p\right)\right), \text{ if } p < \frac{k}{n} < 1.$$

This is proved using the method of types.

We can also change the $(n+1)^2$ in the denominator to $\sqrt{2n}$, by approximating the binomial coefficient with Stirling's formula.

## Bayes' Theorem

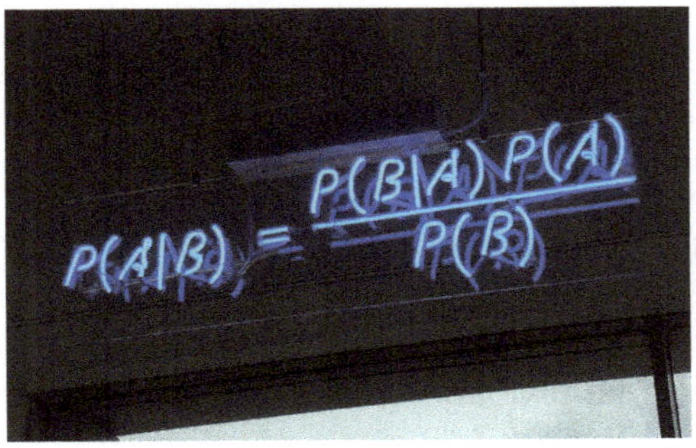


A blue neon sign showing the simple statement of Bayes' theorem at the offices of HP Autonomy.

In probability theory and statistics, Bayes' theorem describes the probability of an event, based on prior knowledge of conditions that might be related to the event. For example, if cancer is related to age, then, using Bayes' theorem, a person's age can be used to more accurately assess the probability that they have cancer than can be done without knowledge of the person's age.

One of the many applications of Bayes' theorem is Bayesian inference, a particular approach to statistical inference. When applied, the probabilities involved in Bayes' theorem may have different probability interpretations. With the Bayesian probability interpretation the theorem expresses how a degree of belief, expressed as a probability, should rationally change to account for availability of related evidence. Bayesian inference is fundamental to Bayesian statistics.

Bayes' theorem is named after Reverend Thomas Bayes, who first used conditional probability to provide an algorithm that uses evidence to calculate limits on an unknown parameter. In what he called a scholium, Bayes extended his algorithm to any unknown prior cause. Independently of Bayes, Pierre-Simon Laplace in 1774, and later in his 1812 "Théorie analytique des probabilités" used conditional probability to formulate the relation of an updated posterior probability from a prior probability, given evidence. Sir Harold Jeffreys put Bayes's algorithm and Laplace's formulation on an axiomatic basis. Jeffreys wrote that Bayes' theorem "is to the theory of probability what the Pythagorean theorem is to geometry."

## Statement of Theorem

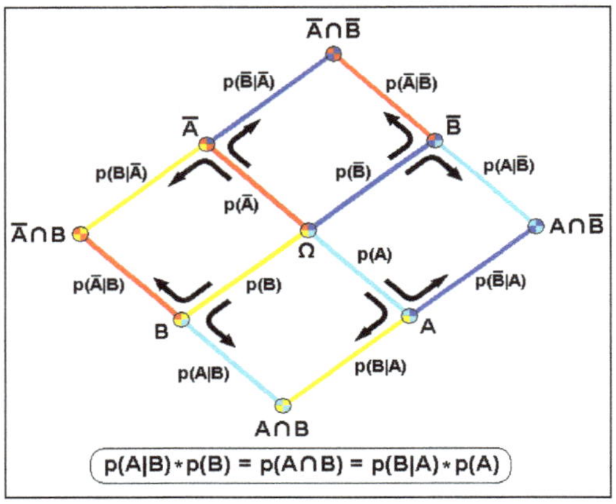


Visualization of Bayes' theorem by superposition of two event tree diagrams.

Bayes' theorem is stated mathematically as the following equation:

$$P(A \mid B) = \frac{P(B \mid A)P(A)}{P(B)}$$

where $A$ and $B$ are events and $P(B) \neq 0$.

- $P(A \mid B)$ is a conditional probability: the likelihood of event $A$ occurring given that $B$ is true.
- $P(B \mid A)$ is also a conditional probability: the likelihood of event $B$ occurring given that $A$ is true.

- $P(A)$ and $P(B)$ are the probabilities of observing $A$ and $B$ independently of each other; this is known as the marginal probability.

## Examples

### Drug Testing

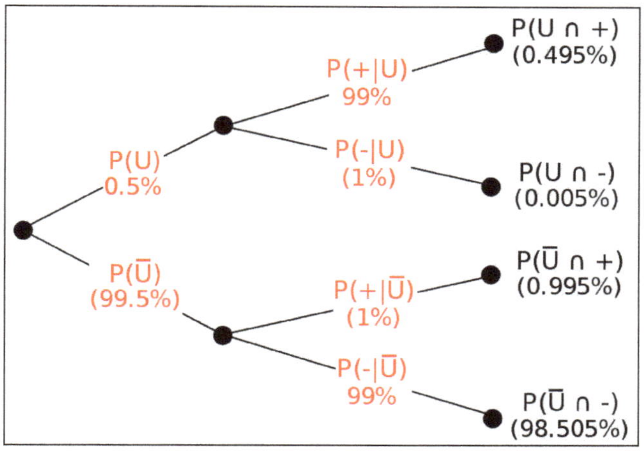


Tree diagram illustrating drug testing example. U, Ū, "+" and "−" are the events representing user, non-user, positive result and negative result. Percentages in parentheses are calculated.

Suppose that a test for using a particular drug is 99% sensitive and 99% specific. That is, the test will produce 99% true positive results for drug users and 99% true negative results for non-drug users. Suppose that 0.5% of people are users of the drug. What is the probability that a randomly selected individual with a positive test is a drug user?

$$\begin{aligned} P(\text{User} \mid +) &= \frac{P(+ \mid \text{User})P(\text{User})}{P(+)} \\ &= \frac{P(+ \mid \text{User})P(\text{User})}{P(+ \mid \text{User})P(\text{User}) + P(+ \mid \text{Non-user})P(\text{Non-user})} \\ &= \frac{0.99 \times 0.005}{0.99 \times 0.005 + 0.01 \times 0.995} \\ &\approx 33.2\% \end{aligned}$$

Even if an individual tests positive, it is more likely that they do not use the drug than that they do. This is because the number of non-users is large compared to the number of users. The number of false positives outweighs the number of true positives. For example, if 1000 individuals are tested, there are expected to be 995 non-users and 5 users. From the 995 non-users, $0.01 \times 995 \simeq 10$ false positives are expected. From the 5 users, $0.99 \times 5 \approx 5$ true positives are expected. Out of 15 positive results, only 5 are genuine.

The importance of specificity in this example can be seen by calculating that even if sensitivity is raised to 100% and specificity remains at 99% then the probability of the person being a drug user only rises from 33.2% to 33.4%, but if the sensitivity is held at 99% and the specificity is increased to 99.5% then the probability of the person being a drug user rises to about 49.9%.

## A More Complicated Example

The entire output of a factory is produced on three machines. The three machines account for 20%, 30%, and 50% of the factory output. The fraction of defective items produced is 5% for the first machine; 3% for the second machine; and 1% for the third machine. If an item is chosen at random from the total output and is found to be defective, what is the probability that it was produced by the third machine?

Once again, the answer can be reached without recourse to the formula by applying the conditions to any hypothetical number of cases. For example, if 100,000 items are produced by the factory, 20,000 will be produced by Machine A, 30,000 by Machine B, and 50,000 by Machine C. Machine A will produce 1000 defective items, Machine B 900, and Machine C 500. Of the total 2400 defective items, only 500 (or 5/24) were produced by Machine C.

A solution is as follows. Let $X_i$ denote the event that a randomly chosen item was made by the $i$ th machine (for $i$ = A,B,C). Let $Y$ denote the event that a randomly chosen item is defective. Then, we are given the following information:

$$P(X_A) = 0.2, \quad P(X_B) = 0.3, \quad P(X_C) = 0.5.$$

If the item was made by the first machine, then the probability that it is defective is 0.05; that is, $P(Y | X_A) = 0.05$. Overall, we have

$$P(Y \mid X_A) = 0.05, \quad P(Y \mid X_B) = 0.03, \quad P(Y \mid X_C) = 0.01.$$

To answer the original question, we first find $P(Y)$. That can be done in the following way:

$$P(Y) = \sum_i P(Y \mid X_i) P(X_i) = (0.05)(0.2) + (0.03)(0.3) + (0.01)(0.5) = 0.024.$$

Hence 2.4% of the total output of the factory is defective.

We are given that $Y$ has occurred, and we want to calculate the conditional probability of $X_C$. By Bayes' theorem,

$$P(X_C \mid Y) = \frac{P(Y \mid X_C) P(X_C)}{P(Y)} = \frac{0.01 \cdot 0.50}{0.024} = \frac{5}{24}$$

Given that the item is defective, the probability that it was made by the third machine is only 5/24. Although machine C produces half of the total output, it produces a much smaller fraction of the defective items. Hence the knowledge that the item selected was defective enables us to replace the prior probability $P(X_C) = 1/2$ by the smaller posterior probability $P(X_C | Y) = 5/24$.

## Interpretations

The interpretation of Bayes' theorem depends on the interpretation of probability ascribed to the terms. The two main interpretations are described.

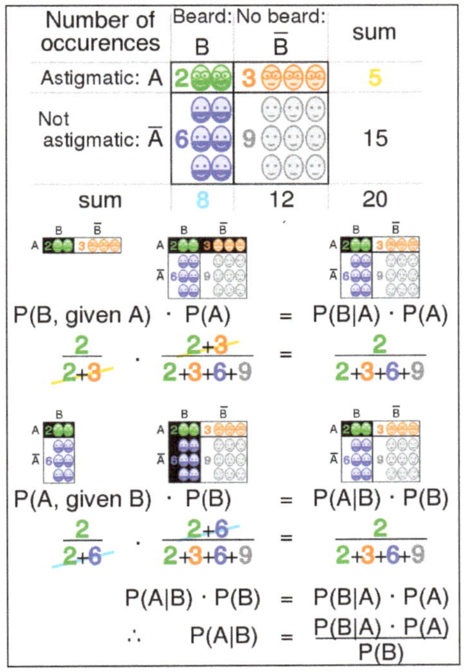


A geometric visualisation of Bayes' theorem. The figures denote the cells of the table involved in each metric, the probability being the fraction of each figure that is shaded. Similar reasoning shows that $P(\bar{A} \mid B) = \frac{P(B|\bar{A})P(\bar{A})}{P(B)}$ and so on.

## Bayesian Interpretation

In the Bayesian (or epistemological) interpretation, probability measures a "degree of belief." Bayes' theorem then links the degree of belief in a proposition before and after accounting for evidence. For example, suppose it is believed with 50% certainty that a coin is twice as likely to land heads than tails. If the coin is flipped a number of times and the outcomes observed, that degree of belief may rise, fall or remain the same depending on the results.

For proposition $A$ and evidence $B$,

- $P(A)$, the *prior*, is the initial degree of belief in $A$.
- $P(A \mid B)$, the *posterior* is the degree of belief having accounted for $B$.
- the quotient $P(B \mid A)/P(B)$ represents the support $B$ provides for $A$.

## Frequentist Interpretation

In the frequentist interpretation, probability measures a "proportion of outcomes." For example, suppose an experiment is performed many times. $P(A)$ is the proportion of outcomes with property $A$, and $P(B)$ that with property $B$. $P(B \mid A)$ is the proportion of outcomes with property $B$ *out of* outcomes with property $A$, and $P(A \mid B)$ the proportion of those with $A$ *out of* those with $B$.

The role of Bayes' theorem is best visualized with tree diagrams, as shown to the right. The two diagrams partition the same outcomes by $A$ and $B$ in opposite orders, to obtain the inverse probabilities. Bayes' theorem serves as the link between these different partitionings.

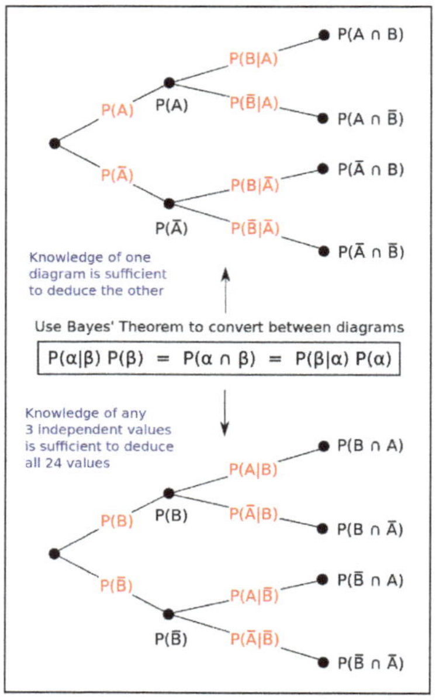


Illustration of frequentist interpretation with tree diagrams. Bayes' theorem connects conditional probabilities to their inverses.

## Example

An entomologist spots what might be a rare subspecies of beetle, due to the pattern on its back. In the rare subspecies, 98% have the pattern, or *P*(Pattern | Rare) = 98%. In the common subspecies, 5% have the pattern. The rare subspecies accounts for only 0.1% of the population. How likely is the beetle having the pattern to be rare, or what is *P*(Rare | Pattern)?

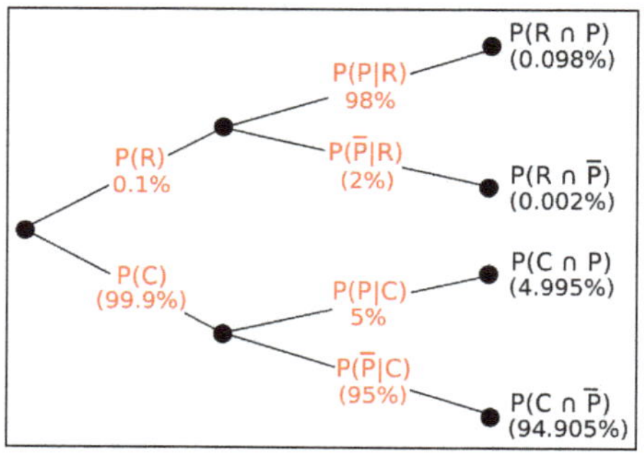


Tree diagram illustrating frequentist example. R, C, P and P bar are the events representing rare, common, pattern and no pattern. Percentages in parentheses are calculated. Three independent values are given, so it is possible to calculate the inverse tree.

From the extended form of Bayes' theorem (since any beetle can be only rare or common),

$$P(\text{Rare} \mid \text{Pattern}) = \frac{P(\text{Pattern} \mid \text{Rare})P(\text{Rare})}{P(\text{Pattern})}$$

$$= \frac{P(\text{Pattern} \mid \text{Rare})P(\text{Rare})}{P(\text{Pattern} \mid \text{Rare})P(\text{Rare}) + P(\text{Pattern} \mid \text{Common})P(\text{Common})}$$

$$= \frac{0.98 \times 0.001}{0.98 \times 0.001 + 0.05 \times 0.999}$$

$$\approx 1.9\%$$

## Forms

### Simple Form

For events $A$ and $B$, provided that $P(B) \neq 0$,

$$P(A \mid B) = \frac{P(B \mid A)P(A)}{P(B)}.$$

In many applications, for instance in Bayesian inference, the event $B$ is fixed in the discussion, and we wish to consider the impact of its having been observed on our belief in various possible events $A$. In such a situation the denominator of the last expression, the probability of the given evidence $B$, is fixed; what we want to vary is $A$. Bayes' theorem then shows that the posterior probabilities are proportional to the numerator:

$P(A \mid B) \propto P(A) \cdot P(B \mid A)$ (proportionality over $A$ for given $B$).

The posterior is proportional to the prior times the likelihood.

If events $A_1, A_2, \ldots,$ are mutually exclusive and exhaustive, i.e., one of them is certain to occur but no two can occur together, and we know their probabilities up to proportionality, then we can determine the proportionality constant by using the fact that their probabilities must add up to one. For instance, for a given event $A$, the event $A$ itself and its complement $\neg A$ are exclusive and exhaustive. Denoting the constant of proportionality by $c$ we have

$$P(A \mid B) = c \cdot P(A) \cdot P(B \mid A) \text{ and } P(\neg A \mid B) = c \cdot P(\neg A) \cdot P(B \mid \neg A).$$

Adding these two formulas we deduce that

$$1 = c \cdot (P(B \mid A) \cdot P(A) + P(B \mid \neg A) \cdot P(\neg A)),$$

or

$$c = \frac{1}{P(B \mid A) \cdot P(A) + P(B \mid \neg A) \cdot P(\neg A)} = \frac{1}{P(B)}.$$

### Alternative Form

Another form of Bayes' theorem that is generally encountered when looking at two competing statements or hypotheses is:

$$P(A \mid B) = \frac{P(B \mid A)P(A)}{P(B \mid A)P(A) + P(B \mid \neg A)P(\neg A)}.$$

For an epistemological interpretation:

For proposition $A$ and evidence or background $B$,

- $P(A)$ is the prior probability, is the initial degree of belief in $A$.

- $P(\neg A)$ is the corresponding probability of the initial degree of belief against $A$, where $1 - P(A) = P(\neg A)$
- $P(B \mid A)$ is the conditional probability or likelihood, is the degree of belief in $B$, given that the proposition $A$ is true.
- $P(B \mid \neg A)$ is the conditional probability or likelihood, is the degree of belief in $B$, given that the proposition $A$ is false.
- $P(A \mid B)$ is the posterior probability, is the probability for $A$ after taking into account $B$ for and against $A$.

## Extended Form

Often, for some partition $\{A_j\}$ of the sample space, the event space is given or conceptualized in terms of $P(A_j)$ and $P(B \mid A_j)$. It is then useful to compute $P(B)$ using the law of total probability:

$$P(B) = \sum_j P(B \mid A_j) P(A_j),$$

$$\Rightarrow P(A_i \mid B) = \frac{P(B \mid A_i) P(A_i)}{\sum_j P(B \mid A_j) P(A_j)}.$$

In the special case where $A$ is a binary variable:

$$P(A \mid B) = \frac{P(B \mid A) P(A)}{P(B \mid A) P(A) + P(B \mid \neg A) P(\neg A)}.$$

## Random Variables

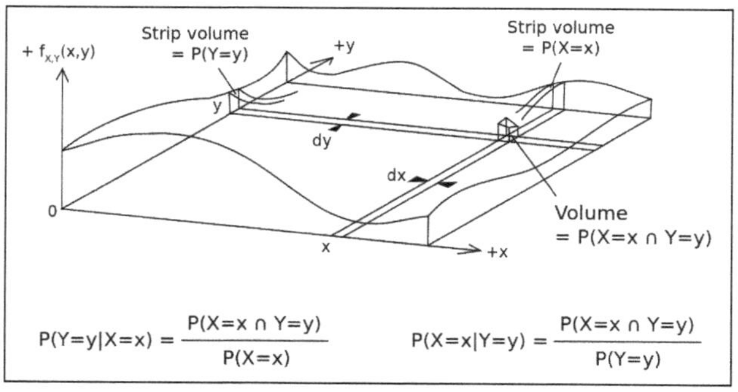


Diagram illustrating the meaning of Bayes' theorem as applied to an event space generated by continuous random variables $X$ and $Y$. There exists an instance of Bayes' theorem for each point in the domain. In practice, these instances might be parametrized by writing the specified probability densities as a function of $x$ and $y$.

Consider a sample space $\Omega$ generated by two random variables $X$ and $Y$. In principle, Bayes' theorem applies to the events $A = \{X = x\}$ and $B = \{Y = y\}$.

$$P(X = x \mid Y = y) = \frac{P(Y = y \mid X = x) P(X = x)}{P(Y = y)}$$

However, terms become 0 at points where either variable has finite probability density. To remain useful, Bayes' theorem may be formulated in terms of the relevant densities.

### Simple Form

If $X$ is continuous and $Y$ is discrete,

$$f_{X|Y=y}(x) = \frac{P(Y=y|X=x)f_X(x)}{P(Y=y)}$$

where each $f$ is a density function.

If $X$ is discrete and $Y$ is continuous,

$$P(X=x|Y=y) = \frac{f_{Y|X=x}(y)P(X=x)}{f_Y(y)}.$$

If both $X$ and $Y$ are continuous,

$$f_{X|Y=y}(x) = \frac{f_{Y|X=x}(y)f_X(x)}{f_Y(y)}.$$

### Extended Form

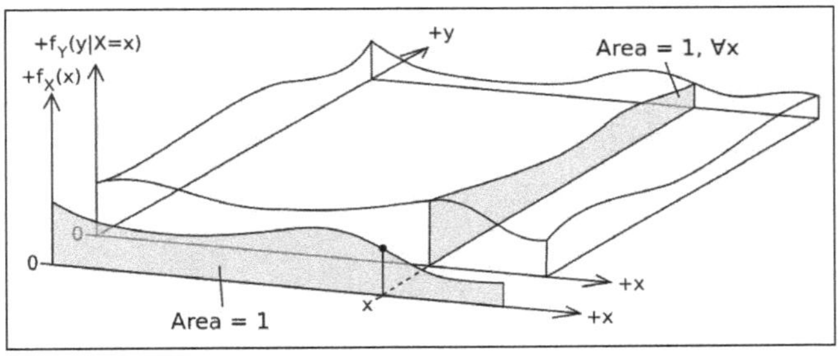


Diagram illustrating how an event space generated by continuous random variables X and Y is often conceptualized.

A continuous event space is often conceptualized in terms of the numerator terms. It is then useful to eliminate the denominator using the law of total probability. For $f_Y(y)$, this becomes an integral:

$$f_Y(y) = \int_{-\infty}^{\infty} f_Y(y|X=\xi)f_X(\xi)d\xi.$$

### Bayes' Rule

Bayes' theorem in odds form is:

$$O(A_1 : A_2 | B) = O(A_1 : A_2) \cdot \Lambda(A_1 : A_2 | B)$$

where

$$\Lambda(A_1 : A_2 | B) = \frac{P(B|A_1)}{P(B|A_2)}$$

is called the Bayes factor or likelihood ratio and the odds between two events is simply the ratio of the probabilities of the two events. Thus

$$O(A_1 : A_2) = \frac{P(A_1)}{P(A_2)},$$

$$O(A_1 : A_2 \mid B) = \frac{P(A_1 \mid B)}{P(A_2 \mid B)},$$

So the rule says that the posterior odds are the prior odds times the Bayes factor, or in other words, posterior is proportional to prior times likelihood.

In the special case that $A_1 = A$ and $A_2 = \neg A$, one writes $O(A) = O(A : \neg A) = P(A)/(1 - P(A))$, and uses a similar abbreviation for the Bayes factor and for the conditional odds. The odds on $A$ is by definition the odds for and against $A$. Bayes' rule can then be written in the abbreviated form

$$O(A \mid B) = O(A) \cdot \Lambda(A \mid B),$$

or in words: the posterior odds on $A$ equals the prior odds on $A$ times the likelihood ratio for $A$ given information $B$. In short, posterior odds equals prior odds times likelihood ratio.

## Derivation

### For Events

Bayes' theorem may be derived from the definition of conditional probability:

$$P(A \mid B) = \frac{P(A \cap B)}{P(B)}, \text{ if } P(B) \neq 0,$$

$$P(B \mid A) = \frac{P(B \cap A)}{P(A)}, \text{ if } P(A) \neq 0,$$

where $P(A \cap B)$ is the joint probability of both A and B being true, because

$$P(B \cap A) = P(A \cap B)$$

$$\Rightarrow P(A \cap B) = P(A \mid B)P(B) = P(B \mid A)P(A)$$

$$\Rightarrow P(A \mid B) = \frac{P(B \mid A)P(A)}{P(B)}, \text{ if } P(B) \neq 0.$$

### For Random Variables

For two continuous random variables *X* and *Y*, Bayes' theorem may be analogously derived from the definition of conditional density:

$$f_{X \mid Y=y}(x) = \frac{f_{X,Y}(x, y)}{f_Y(y)}$$

$$Y \mid X = x\,(y) = \frac{f_{X,Y}(x,y)}{f_X(x)}$$

Therefore,

$$f_{X|Y=y}(x) = \frac{f_{Y|X=x}(y) f_X(x)}{f_Y(y)}.$$

## Correspondence to Other Mathematical Frameworks

### Propositional Logic

*Bayes' theorem* represents a generalisation of contraposition which in propositional logic can be expressed as:

$$(\neg A \to \neg B) \to (B \to A).$$

The corresponding formula in terms of probability calculus is Bayes' theorem which in its expanded form is expressed as:

$$(A \mid B) = \frac{P(B \mid A)a(A)}{P(B \mid A)a(A) + P(B \mid \neg A)a(\neg A)}.$$

In the equation above the conditional probability $P(B \mid A)$ generalizes the logical statement $(A \to B)$, i.e. in addition to assigning TRUE or FALSE we can also assign any probability to the statement. The term $a(A)$ denotes the prior probability (aka. the base rate) of $A$. Assume that $P(A \mid B) = 1$ is equivalent to $(B \to A)$ being TRUE, and that $P(A \mid B) = 0$ is equivalent to $(B \to A)$ being FALSE. It is then easy to see that $P(A \mid B) = 1$ when $P(\neg B \mid \neg A) = 1$ i.e. when $(\neg A \to \neg B)$ is TRUE. This is because $P(B \mid \neg A) = 1 - P(\neg B \mid \neg A) = 0$ so that the fraction on the right-hand side of the equation above is equal to 1, and hence $P(A \mid B) = 1$ which is equivalent to $(B \to A)$ being TRUE. Hence, Bayes' theorem represents a generalization of contraposition.

### Subjective Logic

*Bayes' theorem* represents a special case of conditional inversion in subjective logic expressed as:

$$(\omega^S_{A\tilde{|}B}, \omega^S_{A\tilde{|}\neg B}) = (\omega^S_{B|A}, \omega^S_{B|\neg A}) \tilde{\phi} a_A,$$

where $\tilde{\phi}$ denotes the operator for conditional inversion. The argument $(\omega^S_{B|A}, \omega^S_{B|\neg A})$ denotes a pair of binomial conditional opinions given by source $S$, and the argument $a_A$ denotes the prior probability (aka. the base rate) of $A$. The pair of inverted conditional opinions is denoted $(\omega^S_{A\tilde{|}B}, \omega^S_{A\tilde{|}\neg B})$. The conditional opinion $\omega^S_{A|B}$ generalizes the probabilistic conditional $P(A \mid B)$, i.e. in addition to assigning a probability the source $S$ can assign any subjective opinion to the conditional statement

$(A \mid B)$. A binomial subjective opinion $\omega_A^S$ is the belief in the truth of statement $A$ with degrees of uncertainty, as expressed by source $S$. Every subjective opinion has a corresponding projected probability $P(\omega_A^S)$. The projected probability of opinions applied to Bayes' theorem produces a homomorphism so that Bayes' theorem can be expressed in terms of the projected probabilities of opinions:

$$(\omega_{A\tilde{|}B}^S) = \frac{P(\omega_{B|A}^S)a(A)}{P(\omega_{B|A}^S)a(A) + P(\omega_{B|\neg A}^S)a(\neg A)}$$

Hence, the subjective Bayes' theorem represents a generalization of Bayes' theorem.

## Generalizations

### Conditioned Version

A conditioned version of the Bayes' theorem results from the addition of a third event on which all probabilities are conditioned:

$$P(A \mid B \cap C) = \frac{P(B \mid A \cap C)P(A \mid C)}{P(B \mid C)}$$

### Derivation

Using the chain rule

$$P(A \cap B \cap C) = P(A \mid B \cap C)P(B \mid C)P(C)$$

And, on the other hand

$$P(A \cap B \cap C) = P(B \cap A \cap C) = P(B \mid A \cap C)P(A \mid C)P(C)$$

The desired result is obtained by identifying both expressions and solving for $P(A \mid B \cap C)$.

# Conditional Probability

Conditional probability P(A | B) indicates the probability of event 'A' happening given that event B happened.

$$P(A \mid B) \quad \frac{P(A \cap B)}{P(B)}$$

We can easily understand the above formula using the below diagram. Since B has already happened, the sample space reduces to B. So the probability of A happening becomes $P(A \cap B)$ divided by P(B)

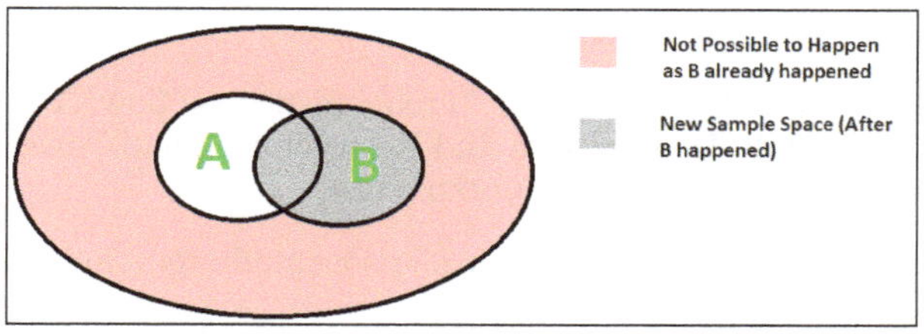


Example: In a batch, there are 80% C programmers, and 40% are Java and C programmers. What is the probability that a C programmer is also Java programmer?

Let A --> Event that a student is Java programmer

B --> Event that a student is C programmer

$P(A|B) = P(A \cap B) / P(B)$

$= (0.4) / (0.8)$

$= 0.5$

So there are 50% chances that student that knows C also knows Java.

## Product Rule

Derived from above definition of conditional probability by multiplying both sides with P(B)

$$P(A \cap B) = P(B) * P(A|B)$$

## Understanding Conditional probability through tree

Computation for Conditional Probability can be done using tree, This method is very handy as well as fast when for many problems.

Example: In a certain library, twenty percent of the fiction books are worn and need replacement. Ten percent of the non-fiction books are worn and need replacement. Forty percent of the library's books are fiction and sixty percent are non-fiction. What is the probability that a book chosen at random are worn? Draw a tree diagram representing the data.

Solution: Let F represents fiction books and N represents non-fiction books. Let W represents worn books and G represents non-worn books.

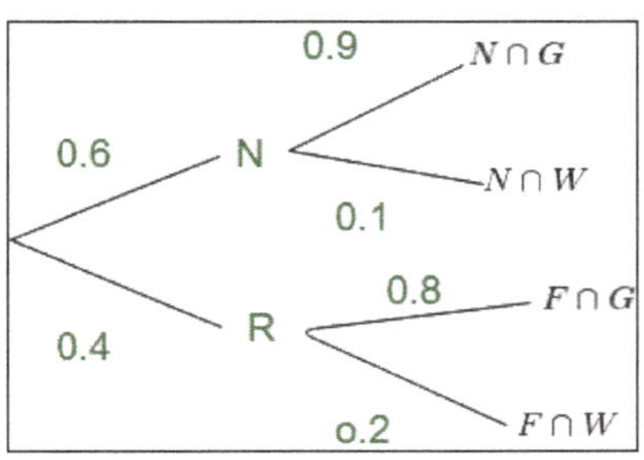


$P(worn) = P(N)*P(N | W) + P(F)*P(F | W)$

$= 0.6*0.1 + 0.4*0.2$

$= 0.14$

## References


- Chen, P.; Chen, Z.; Bak-Jensen, B. (April 2008). "Probabilistic load flow: A review". 2008 Third International Conference on Electric Utility Deregulation and Restructuring and Power Technologies. Pp. 1586–1591. Doi:10.1109/drpt.2008.4523658. ISBN 978-7-900714-13-8
- What-is-probability-in-math-definition-lesson-quiz, lesson: study.com, Retrieved 26 July, 2019
- Olav Kallenberg; Probabilistic Symmetries and Invariance Principles. Springer -Verlag, New York (2005). 510 pp. ISBN 0-387-25115-4
- Probability-of-events, probability-and-statistic, pre-algebra, education: mathplanet.com, Retrieved 21 May, 2019
- Kallenberg, Olav (2002). Foundations of Modern Probability (2nd ed.). New York: Springer. P. 9. ISBN 0-387-94957-7
- Yates, Daniel S.; Moore, David S; Starnes, Daren S. (2003). The Practice of Statistics (2nd ed.). New York: Freeman. ISBN 978-0-7167-4773-4. Archived from the original on 2005-02-09. Retrieved 2013-07-18
- Theorem 11.1.3 in Cover, T.; Thomas, J. (2006). Elements of Information Theory (2nd ed.). Wiley. P. 350. ISBN 9781118585771

# Permissions

All chapters in this book are published with permission under the Creative Commons Attribution Share Alike License or equivalent. Every chapter published in this book has been scrutinized by our experts. Their significance has been extensively debated. The topics covered herein carry significant information for a comprehensive understanding. They may even be implemented as practical applications or may be referred to as a beginning point for further studies.

We would like to thank the editorial team for lending their expertise to make the book truly unique. They have played a crucial role in the development of this book. Without their invaluable contributions this book wouldn't have been possible. They have made vital efforts to compile up to date information on the varied aspects of this subject to make this book a valuable addition to the collection of many professionals and students.

This book was conceptualized with the vision of imparting up-to-date and integrated information in this field. To ensure the same, a matchless editorial board was set up. Every individual on the board went through rigorous rounds of assessment to prove their worth. After which they invested a large part of their time researching and compiling the most relevant data for our readers.

The editorial board has been involved in producing this book since its inception. They have spent rigorous hours researching and exploring the diverse topics which have resulted in the successful publishing of this book. They have passed on their knowledge of decades through this book. To expedite this challenging task, the publisher supported the team at every step. A small team of assistant editors was also appointed to further simplify the editing procedure and attain best results for the readers.

Apart from the editorial board, the designing team has also invested a significant amount of their time in understanding the subject and creating the most relevant covers. They scrutinized every image to scout for the most suitable representation of the subject and create an appropriate cover for the book.

The publishing team has been an ardent support to the editorial, designing and production team. Their endless efforts to recruit the best for this project, has resulted in the accomplishment of this book. They are a veteran in the field of academics and their pool of knowledge is as vast as their experience in printing. Their expertise and guidance has proved useful at every step. Their uncompromising quality standards have made this book an exceptional effort. Their encouragement from time to time has been an inspiration for everyone.

The publisher and the editorial board hope that this book will prove to be a valuable piece of knowledge for students, practitioners and scholars across the globe.

# Index

CPSIA information can be obtained
at www.ICGtesting.com
Printed in the USA
LVHW060859030122
707669LV00002B/262

9 781632 388780